Synthetic Biodegradable Polymer Scaffolds

Anthony Atala
David J. Mooney
Editors

Joseph P. Vacanti
Robert Langer
Associate Editors

Birkhäuser
Boston • Basel • Berlin

R
857
,P6
S957
1997

Anthony Atala
Department of Surgery
Children's Hospital and Harvard Medical School
Boston, MA 02115

Joseph P. Vacanti
Department of Surgery
Children's Hospital
Boston, MA 02115

David J. Mooney
University of Michigan, Ann Arbor
3074 Herbert Dow Building
Ann Arbor, MI 48109

Robert Langer
Department of Chemical Engineering
Massachusetts Institute of Technology
Cambridge, MA 02139

Library of Congress Cataloging-in-Publication Data
Synthetic biodegradable polymer scaffolds / Anthony Atala and David J.
  Mooney, editors : Joseph P. Vacanti and Robert Langer, associate
  editors.
      p.   cm. — (Tissue engineering)
  Includes bibliographical references and index.
  ISBN 0-8176-3919-5 (alk. paper)
  1. Polymers in medicine—Biodegradation. 2. Tissue culture.
3. Animal cell biotechnology. I. Atala, Anthony, 1958–   .
II. Mooney, David J., 1964–   . III. Series.
R857.P6S957   1997
617.9'5—dc21
                                        96-45700
                                        CIP

Printed on acid-free paper
© 1997 Birkhäuser Boston

COVER PHOTOS—Left: Polyglycolic acid polymer scaffold with urothelial cells implanted in vivo and retrieved at 30 days shows cell layering and spatial orientation similar to that of normal urothelium. Reduced from X100. **Center:** Polyglycolic acid polymer without cells. Reduced from X200. **Right:** Formation of composite multilayered sheet-like structures lining tube obtained from cell-polymer implant consisting of urothelial and smooth muscle cells. Reduced from X200. Courtesy of Anthony Atala, Harvard Medical School, Boston.

ISBN 0-8176-3919-5
ISBN 3-7643-3919-5
Typeset by Northeastern Graphic Services, Inc., Hackensack, NJ.
Printed and bound by Hamilton Printing Co., Rennselaer, NY.
Printed in the U.S.A
9 8 7 6 5 4 3 2 1

Tissue Engineering

Series Editor—
Anthony Atala, MD
Children's Hospital, Boston

Editorial Advisory Board—
Jeffrey Hubbell, PhD
California Institute of Technology

Robert S. Langer, ScD
Massachusetts Institute of Technology

Antonios G. Mikos, PhD
Rice University

Joseph P. Vacanti, MD
Children's Hospital, Boston

Forthcoming Titles in the Series—
Encapsulated Cell Technology and Therapeutics
R. Lanza and W. L. Chick, eds.

Wound Healing and Tissue Engineering
J. O. Hollinger and D. M. Toriumi, eds.

List of Contributors

Anthony Atala, MD, Director, Laboratory for Tissue Engineering, Division of Urology and Department of Surgery, Children's Hospital and Harvard Medical School, 300 Longwood Avenue, Boston, MA 02115

Ronnda Bartel, PhD, Advanced Tissue Sciences, Inc., 10933 N. Torrey Pines Road, La Jolla, CA 92037

Rebekah D. Bostrom, BA, Cox Laboratory for Biomedical Engineering, Institute of Biosciences and Bioengineering and Department of Chemical Engineering, Rice University, P.O. Box 1892, Houston, TX 77251-1892

Linda G. Braddon, PhD, Institute for Bioengineering and Bioscience, Georgia Institute of Technology, Atlanta, GA 30332-0363

Yilin Cao, MD, PhD, Laboratory for Tissue Engineering, Department of Anesthesiology, University of Massachusetts Medical Center, 55 Lake Avenue North, S2-751, Worcester, MA 01655-0300

Beverly E. Chaignaud, MD, Department of Surgery, Enders 1161, 320 Longwood Avenue, Children's Hospital, Boston, MA 02115

Jeffrey A. Hubbell, PhD, California Institute of Technology, Mail Code 210-41, Pasadena, CA 91125

Clemente Ibarra, MD, Laboratory for Tissue Engineering, Department of Anesthesiology, University of Massachusetts Medical Center, 55 Lake Avenue North, S2-751, Worcester, MA 01655-0300

William H. Kitchens, JD, Arnall Golden & Gregory, 2800 One Atlantic Center, 1201 West Peachtree Street, Atlanta, GA, 30309-3400

Robert Langer, ScD, Department of Chemical Engineering, Massachusetts Institute of Technology, Room B25-342, 77 Massachusetts Avenue, Cambridge, MA 02139

Hanmin Lee, MD, Department of Surgical Research, 1155 Enders, Children's Hospital, 320 Longwood Avenue, Boston, MA 02115

Kam W. Leong, PhD, Department of Biomedical Engineering, The Johns Hopkins University, School of Medicine, Baltimore, MD 21205

Jonathan Mansbridge, PhD, Advanced Tissue Sciences, Inc., 10933 N. Torrey Pines Road, La Jolla, CA 92037

John E. Mayer Jr., MD, Department of Cardiac Surgery, Children's Hospital, 300 Longwood Avenue, Boston, MA 02115

Antonios G. Mikos, PhD, T.N. Law Associate Professor, Cox Laboratory for Biomedical Engineering, Institute of Biosciences and Bioengineering and Department of Chemical Engineering, Rice University, P.O. Box 1892, Houston, TX 77251-1892

David J. Mooney, PhD, Departments of Biologic and Materials Sciences and Chemical Engineering, The University of Michigan, Ann Arbor, MI 48109

Gail K. Naughton, PhD, Advanced Tissue Sciences Inc., 10933 N. Torrey Pines Road, La Jolla, CA 92037

Robert M. Nerem, PhD, Institute for Bioengineering and Bioscience, Georgia Institute of Technology, Atlanta, GA 30332-0363

Patrea L. Pabst M.S., JD, Arnall Golden & Gregory LLP, 2800 One Atlantic Center, 1201 West Peachtree Street, Atlanta, GA 30309-3450

Dror Seliktar, MS, Institute for Bioengineering and Bioscience, Georgia Institute of Technology, Atlanta, GA 30332-0363

Toshiharu Shin'oka, MD, Department of Surgical Research, Enders 1155, 320 Longwood Avenue, Children's Hospital, Boston, MA 02115

Charles A. Vacanti, MD, Laboratory for Tissue Engineering, Department of Anesthesiology, University of Massachusetts Medical Center, 55 Lake Avenue North, S2-751, Worcester, MA 01655-0300

Joseph P. Vacanti, MD, Department of Surgery, Enders 1161, 320 Longwood Avenue, Children's Hospital, Boston, MA 02115

Jennifer L. West, PhD, Department of Bioengineering, Rice University, Houston, TX 77251-1892

Wai Hung Wong, PhD, Departments of Biologic and Materials Sciences and Chemical Engineering, The University of Michigan, Ann Arbor, MI 48109

Thierry Ziegler, PhD, Institute for Bioengineering and Bioscience, Georgia Institute of Technology, Atlanta, GA 30332-0363

Contents

Series Preface

This body of work represents the first volume of a book series covering the field of tissue engineering. Tissue engineering, which refers to a category of therapeutic or diagnostic products and processes which are based upon a combination of living cells and biomaterials, was defined as a field only a few years ago (1988). Tissue engineering is an inherently interdisciplinary field, combining bioengineering, life sciences and clinical sciences. The definition of this area of work as the field of tissue engineering brought together scientists from multiple backgrounds who already were working toward the achievement of similar goals.

Why a book series exclusively devoted to tissue engineering? The field of tissue engineering is heterogeneous. The cells involved in tissue engineering can be autologous, allogeneic or xenogeneic. The biomaterials utilized can be either naturally occurring, synthetic or a combination of both. The application of the technology can be either for acute or permanent purposes. An attempt to cover the field of tissue engineering in a single volume, with the degree of detail necessary for individuals with different scientific backgrounds and disciplines, would be a difficult task to accomplish, particularly when this field is just emerging and changing rapidly. Therefore, addressing different technologies within the field of tissue engineering, in a comprehensive manner, is the main mission of this series of volumes. A stellar group of scientists has been brought together to form the editorial board of the series. The distinguished members of the editorial board are some of the current leaders in the field, and all share the same goals of making this series of volumes the "Bible" in regards to all technologies within the field of tissue engineering.

The first volume of the series is dedicated to the technology involving synthetic biodegradable polymers as extracellular matrices. This area was chosen as the subject of the first volume due to the wide applicability of the technology to many areas in the clinical sciences. Today, health care costs for millions of Americans who suffer tissue loss or end-stage organ failure exceed $400 billion per year. Tissue engineering holds the promise of delivering new and less expensive solutions to the medical field. Although the field is young, multiple clinical trials are in progress already. Certainly within the area of tissue engineering, what was considered science fiction only a few years ago has now become a reality. Thus, this series is dedicated to engineers, life scientists, clinicians and other individuals involved in tissue engineering, both in industry and academia, who are striving to advance the field for the benefit of mankind.

Anthony Atala, M.D.
Series Editor

Preface

Broadly described, the emerging field of tissue engineering includes studies in areas as diverse as synthesis of new polymers, studies of signal transduction and gene regulation in cultured cells, and immunology issues related to transplantation. This book does not aspire to cover all of the diverse issues relevant to tissue engineering; rather, it focuses on approaches to engineering new tissues using synthetic biodegradable polymers as extracellular matrices.

These synthetic matrices are utilized to: 1) deliver transplanted cells to a desired anatomic location and/or control the ingrowth of host cells into the new tissue, 2) create potential space for tissue development, and 3) guide the process of tissue formation. Both naturally-derived (e.g., collagen) and synthetic polymers (e.g., polylactic acid) have been utilized to fabricate these matrices. Naturally-derived materials must be isolated from human, animal, or plant tissue, which typically results in high costs and large batch-to-batch variations. In addition, these materials typically exhibit a very limited range of properties and are often difficult to process. Synthetic polymers, in contrast, are both cheaply produced and reproducible, and they can be readily processed into a variety of matrices with desirable macro- and microstructures. Biodegradable polymers are attractive materials because they can be designed to erode once they accomplish their function, resulting in the formation of a completely natural tissue with no permanent synthetic element. Thus, many difficulties commonly associated with the permanent implantation of synthetic materials (e.g., infection and chronic inflammation) may be avoided.

This first volume in the *Tissue Engineering* series was developed not only to bring together an accessible collection of polymer, cellular, and tissue studies, but also to introduce important legal issues in engineering tissues with synthetic biodegradable polymers. Included are chapters on the synthesis and properties of these polymers, their application to engineering a variety of tissue types, and relevant FDA and patent issues. This text will serve as a useful reference to researchers and clinicians already involved in the field as they seek guidance on potential new materials or applications, and to students or beginning researchers who need an overview of the field.

This book begins with a history of tissue engineering in order to place the current state of the field in a historical perspective. The introductory chapters (Chapters 1–3) present this history and discuss legal issues which will be critical to this emerging industry. A valuable overview of patent and other intellectual property issues leads into a discussion of FDA regulatory strategies. The next section (Chapters 4–5) examines the fundamentals of biodegradable polymer scaffolds: Chapter 4 reviews the synthesis and properties of those synthetic polymers presently utilized in the field, as well as the numerous other polymers that could be useful in the future; Chapter 5 examines a critical area of research—the synthesis of biologically active polymers. The last section (Chapters 6-13) details the wide variety of applications for these polymers in different systems. Chapter 6 describes how these polymers can be utilized for drug delivery, and the remaining chapters describe engineering in epithelial tissue, urologic tissue, vascular structures, cartilage, bone, and liver.

We acknowledge the authors for their outstanding efforts in achieving concise, yet comprehensive, overviews of their topics.

David J. Mooney and Anthony Atala

Note added in proof:
Regarding Chapter 3—FDA Regulation of Tissue Engineering
As this book went to press, the U.S. Food and Drug Administration on February 28, 1997 proposed a new regulatory framework for human cellular and tissue-based products. The proposal provides a tiered approach to regulation with increasing degrees of government oversight as the potential risk increases. Tissues that do not undergo extensive processing will be subject to infectious disease screening and testing and to requirements aimed at preventing contamination and preserving integrity and function, but will not need FDA review or marketing approval. FDA, however, will require approval of human testing and premarket approval based on a demonstration of safety and effectiveness for tissues that are processed extensively (i.e., their biological or functional characteristics have been manipulated), used for purposes other than their normal function, or combined with devices, drugs or other biologics. Technologies such as somatic cell therapy and gene therapy will be in this category as will stem cell therapy in patients not closely related to the cell donor. Because this proposal supplements and in some instances deviates from previous guidance that was available at the time this chapter was written, the reader is encouraged to review this new FDA proposal as well.

William H. Kitchens

1

THE HISTORY OF TISSUE ENGINEERING USING SYNTHETIC BIODEGRADABLE POLYMER SCAFFOLDS AND CELLS

BEVERLY E. CHAIGNAUD, ROBERT LANGER, AND JOSEPH P. VACANTI

INTRODUCTION

In the not too distant past, organ transplantation was considered a futuristic concept. However, things considered miraculous in one era may be merely remarkable in another (Langer and Vacanti, 1995). Over the last half of this century, development of techniques for tissue and organ transplantation have been revolutionary. Along with the tremendous advances in the field of transplantation, new problems have emerged needing solutions. It is in this context that the field of tissue engineering has emerged over the last decade (Vacanti and Vacanti, 1996).

TISSUE ENGINEERING DEFINED

Tissue engineering is "an interdisciplinary field that applies principles of engineering and the life sciences toward the development of biological substitutes that restore, maintain, and improve the function of damaged tissues and organs" (Skalak and Fox, 1988; Langer and Vacanti, 1993; Nerem, 1991). As a multidisciplinary science, tissue engineering merges the fields of cell biology, engineering, materials science, and surgery to fabricate new functional tissue using living cells and a matrix or scaffolding which can be natural, man-made, or a composite of both. Defined as a field less than ten

Synthetic Biodegradable Polymer Scaffolds
Anthony Atala and David Mooney, Editors
Robert Langer & Joseph P. Vacanti, Associate Editors
© 1997 Birkhäuser Boston

years ago, tissue engineering remains in its embryonic stage of development. As in the early stages of any new field, much of the progress to date has been related to the development of model systems and the delineation of principles of development (Vacanti and Vacanti, 1996).

Three general strategies have been utilized in the creation of new tissue (Langer and Vacanti, 1993). These include: (1) the replacement of only those isolated cells or cell substitutes needed for function; (2) production and delivery of tissue-inducing substances such as growth factors and signal molecules; (3) cells placed on or within a matrix fashioned from synthetic polymers or natural substances such as collagen. The matrix system may be an open or closed system. Open cell-matrix systems are implanted in the body and become completely integrated with the host tissue, allowing free transport of molecules and cells between the host tissue and implanted cells. In a closed matrix system, cells are isolated from the body by a membrane that permits nutrient and gas exchange while acting as a barrier for large entities such as antibodies and immune cells. Closed matrix systems may be implanted in the body or used as extracorporeal devices (Langer and Vacanti, 1993; Mooney and Langer, 1995). We will focus here on the third of these strategies of tissue engineering, the creation of new tissue using living cells and synthetic biodegradable polymer matrices.

TISSUE ENGINEERING USING BIODEGRADABLE POLYMERS

Over the past decade, our laboratory has been involved in research to generate new, natural, permanent tissue replacements by creating implantable devices composed of tissue-specific cells on synthetic biodegradable polymer scaffolds. This system offers the advantage of being able to use either allogenic cells or autologous transplants (Vacanti, 1993). The manmade scaffolds provide a structural framework for the selected cells and facilitate the formation of new tissues. The open, porous, three-dimensional structure of the degradable polymers has been designed to maximize diffusion parameters and to permit vascular ingrowth into the implanted structures (Mooney et al, 1992; Vacanti, 1993). The polymer is completely resorbed over time leaving only the newly formed tissue, a chimera of donor cells and recipient mesenchymal elements including blood vessels and connective tissue (Mooney et al, 1992).

Origin of concept

Developments in medical science that led to formation of the concepts of tissue engineering using biodegradable polymer scaffolds and cells occurred in Boston. In 1954, at Peter Bent Brigham Hospital in Boston, a team headed by Joseph Murray, MD, performed the first successful solid organ transplant (Murray et al, 1955). They transplanted a kidney from one identical twin into his critically ill brother. It was again under Dr. Murray's direction that the transplantation of tissue from one nongenetically identical individual into another was carried out in the early 1960s at Peter Bent Brigham Hospital. These singular events followed years of research in the laboratory and opened the door to the fields of solid organ transplantation, transplantation biology, and immunology that exist today (Vacanti, 1988).

The transplantation of extrarenal solid organs remained largely experimental until the late 1970s. With the development of improved immunosuppression to control rejection of the foreign tissue and advancements in the creation of standardized surgical and anesthetic techniques, solid organ transplantation underwent explosive growth (Vacanti, 1988).

While tremendous strides have been made with the advancing technology of transplantation since its inception over forty years ago, many inherent problems still exist and new problems have emerged. The dominant problem facing transplantation today is tissue and organ shortage. As of December 1995, there were over 44,000 people on waiting lists for organs in the United States, and many of these people will die before an organ becomes available (UNOS Update, 1996). Experimental approaches to solve the problem of organ shortage have included xenotransplantation and the development of bioartificial devices. Surgical advances to address the organ need include segmental liver transplants and living related liver transplants. While these approaches have been beneficial, the problem of organ scarcity remains. Other problems surrounding organ transplantation include complex labor intensive care, the expense of transplantation, and both the short and long term problems of immunosuppression including rejection and tumor formation. These shortcomings have been recognized for years, and researchers have looked beyond transplantation for solutions (Vacanti, 1988).

In 1985, Paul S. Russell MD, wrote a review article on selective cell transplantation summarizing new approaches to organ transplantation (Russell, 1985). The concept of selective transplantation described by Russell referred to the selection of only a part of an organ or tissue for transplantation. Islet cell transplantation and transfer of dissociated liver cells were two areas he discussed as conceptually offering potential for the restoration of function. While isolated cells are capable of forming tissue structures when implanted, this process is limited when the cells are placed as a cell suspension into tissue. Implanted in this manner, the cells lack any intrinsic beginning structure, and they lack a template to guide restructuring. The volume of tissue that can be successfully implanted is also limited by the fact that cells must be no more than a few hundred microns from a capillary to survive (Folkman and Hochberg, 1973; Vacanti, 1993).

The pediatric liver transplantation program at Children's Hospital in Boston began in 1984. There were six centers for liver transplantation in the United States at that time. Research work at other transplant centers was involved in understanding and suppressing the immune system. The problem of organ scarcity was being addressed scientifically with the use of xenografts or biohybrid temporary support devices. No one was addressing the problem of organ shortage from the standpoint of fabricating functional replacement tissues and organs. It was here that the research of the newly founded Laboratory of Tissue Engineering and Transplantation at Children's Hospital in Boston under Vacanti's direction and the polymer science laboratory of Professor Robert Langer at Massachusetts Institute of Technology would focus.

Knowing the work of John F. Burke, MD and Ioannis, Yannas, MD using a two-dimensional artificial matrix to create a neodermis and the work of Russell with cell transplantation, Vacanti wondered if the same principles could be applied to a three dimensional system to make organs. Beginning a close collaboration that continues today, the senior authors applied principles of cell transplantation and synthetic biodegradable polymer matrices to the development of a three-dimensional system to create visceral organs (Vacanti, 1988). This work began in 1985.

The use of a three-dimensional system to create new tissues and organs by attaching parenchymal cells to biodegradable polymer scaffolds was based on several biological observations (Vacanti, 1988). (1) All organisms undergo continual renewal, remodeling, and replacement. (2) Dissociated cells will reform structures when provided with cues from their environment as demonstrated by the formation of tubular structures by endothelial cells and ductal structures by biliary cells (Folkman and Haudenschild, 1980; Morse and Vacanti, 1988; Vacanti, 1988). (3) Organ parenchymal cells are anchorage dependent with the cell matrix playing an important role in cell shape, division, differentiation, and function (Folkman and Moscona, 1978; Ben-Ze'ev et al, 1980; Vacanti, 1988; Vacanti et al, 1988). (4) Two to three mm^3 represents the largest volume of tissue that can be implanted in order to achieve nutrition, gas exchange, and removal of waste products until angiogenesis occurs (Folkman and Hochberg, 1973; Vacanti, 1988).

The three-dimensional system as envisioned would require parenchymal cells and a biodegradable scaffold that would permit diffusion of nutrients, promote vascular ingrowth, and permit cellular proliferation and function. Functional dissociated cells would be seeded onto the biodegradable polymer scaffold in culture and the cell-polymer constructs would then be implanted into a host (Vacanti, 1988; Vacanti and Vacanti, 1991). It was reasoned that attachment of cells to an organized scaffold prior to implantation would permit implantation of an increased number of cells (Vacanti et al, 1988).

The ideal matrix

Certain criteria were considered necessary for an ideal matrix for cell transplantation. The matrix should be biocompatible, not inducing a tissue response in the host, and completely resorbable leaving a totally natural tissue replacement following degradation of the polymer. The matrix should be easily and reliably reproducible into a variety of shapes and structures that retain their shape when implanted. As a vehicle for cell delivery, the matrix should provide mechanical support to maintain a space for tissue to form (Mooney and Langer, 1995). The interaction of the surface of the matrix with cells should support differentiated cell function and growth (Cima et al, 1991b), and in certain situations should induce ingrowth of desirable cell types from surrounding tissue.

Polymers in the group of polyesters, specifically the family of polylactic acid (PLA) and polyglycolic acid (PGA) and copolymers of lactic and glycolic acids (PLGAs), most closely fulfilled the criteria outlined above including biocompatibility, processibility, and controlled degradation (Cima et al, 1991b). These polymers, many descendants of absorbable suture materials developed two decades ago (Langer and Vacanti, 1995), were approved for in vivo use by the FDA for certain applications and were readily processible into a variety of shapes and forms using melt and solvent techniques. (Cima et al, 1991b; Langer and Vacanti, 1995). The polymers degraded by hydrolysis leaving natural metabolic intermediates, and the resorption rates could be designed to vary from months to years depending on the ratio of the monomers (Cima et al, 1991b). In addition, the polymers could potentially be manufactured to provide controlled release of hormones and growth factors (Cima et al, 1991a).

These polymers were selected for the first matrix designs used in our laboratory with the intention of modifying the surface chemistry as necessary to provide optimal cell-polymer interactions (Cima et al, 1991b). The optimal matrix microstructure was to some extent determined by the tissue to be transplanted. However, certain parameters pertained to all matrices for cell transplantation. Matrix structures should be open with high porosity to provide diffusion of nutrients, and there should be a large surface area to volume ratio to provide a large surface area for cell-polymer interactions (Cima et al, 1991b). The variety of processing techniques available was a distinct advantage in using synthetic polymers. Fibers, hollow fibers, porous sponges, and any number of designs could be readily formed (Mooney and Langer, 1995).

From concept to reality

Progression from concept to experiments moved quickly. The first polymer design consisted of a small wafer of biodegradable polyanhydride. Hepatocytes were seeded in a monolayer on the polymer disc and implanted into a recipient animal. From analysis of these initial experiments using cells placed on the polymer matrices it was found that cell number and cell density were inadequate for reliable successful engraftment (Vacanti, 1988; Vacanti and Vacanti, 1991). This led to questions of growth of multicellular organisms, and in this context it was asked how nature solves the problem of three-dimensional growth. How does nature "make the inside the outside" so that cells on the interior can be nourished (Vacanti, 1988).

Answers to complex questions are often found in familiar objects that surround us. Nature's solution to three-dimensional growth was found in the fronds of seaweed on a Cape Cod shore. It is known that the surface area of an enlarging mass of cells increases as the square of the radius, while the volume increases as a cube of the radius (Vacanti, 1988; Vacanti and Vacanti, 1991). Branching networks are nature's solution to matching surface area to volume. Just as in the simple branching pattern of seaweed, organs in both the plant and animal worlds are composed of and communicate through intertwined branching networks (MacDonald, 1983; Vacanti, 1988). We therefore hypothesized that open branching networks of polymers trying to match surface area to volume may keep large masses of cells alive.

To test this concept, three-dimensional, synthetic, biodegradable polymers of branching fiber networks were formed and seeded with parenchymal cell suspensions from the liver, intestine, and pancreas and implanted (Vacanti et al, 1988). Aortic endothelial cells were also seeded on polymer and suspended in a gel biomatrix, and proliferation of cells and migration of cells off the matrix reminiscent of capillary branching was observed (Vacanti, 1988).

Three types of synthetic absorbable polymers were used in these first experiments. Wafer discs or filaments of polyanhydrides and polyorthoesters were fabricated using techniques of solvent casting, compression molding, or filament drawing. Branching pattern relief structures were etched into the discs after formation mimicking the branching patterns of nature. Fibers of polyglactin 910 converging to a common base were fashioned from this polymer which was developed as absorbable suture material (Vacanti et al, 1988).

Standard techniques of cell harvest were used, and single cell and clusters of the fetal and adult rat and mouse hepatocytes, pancreatic islet cells, and small intestine cells were seeded onto the polymer matrices. Most of the host animals underwent a partial hepatectomy to stimulate cell growth. The cell-polymer constructs were maintained in culture four days and then implanted into host animals in one of three sites, the bowel mesentery, the omentum, or the interscapular fat pad (Vacanti et al, 1988).

Results of these initial experiments using cells and biodegradable polymer scaffolds demonstrated that cells from liver, pancreas, and intestine could be harvested and that the cells would attach to the synthetic, biodegradable polymer matrix. The cells were noted to survive in culture and could be implanted on the polymer. Successful engraftment of small clusters of hepatocytes and intestinal cells in a number of the cell-polymer constructs was demonstrated. Histology of the explanted tissue showed small clusters of healthy appearing hepatocytes with bile canaliculi and mitotic figures in some areas. Blood vessel ingrowth was observed three days after implantation with all the polymer types, and new blood vessels were noted in the interstices between the polymer filaments following implantation. Cell viability depended on the polymer type and ranged from less than 10% on the polyahydride discs to 80% on the polyorthoester discs and filaments, and over 90% cell viability on polyglactin 910. Evidence of both acute and chronic inflammatory responses were also observed and varied with the polymer used with polyanhydrid eliciting the most intense response. Following these first experiments, further studies were needed to define optimal polymer characteristics, attachment parameters, cell growth criteria, and cell function (Vacanti et al, 1988). However, a novel approach to cell and tissue transplantation had been introduced.

This concept of engineering a new organ in situ by placing dissociated cells onto synthetic biodegradable polymer scaffolds in culture and then placement of the cell-polymer construct into a host to permit vascularization, growth, and function was termed *chimeric neomorphogenesis* (Vacanti, 1988). The authors reasoned that if functional, the newly formed tissue would be a true chimera consisting of parenchymal elements of the donor and mesenchymal elements of the recipient (Vacanti et al, 1988). Embryogenesis in the field of tissue engineering using synthetic biodegradable polymers and cells had begun.

Model systems include liver and a variety of tissues

From its inception, much of the work in our laboratories has centered on the engineering of liver tissue, investigating the growth and function of transplanted hepatocyes in vitro as well as in small and large animal models.

Working to develop model systems, our research focus rapidly expanded to include a variety of different tissues and organs. While many of the issues of cell transplantation are the same for all cell types, unique characteristics and inherent advantages and difficulties exist for each different cell type. Correct conditions for cell survival, differentiation, and growth must be provided for each different cell type (Vacanti and Vacanti, 1991). Much work in our laboratories has gone into understanding and optimizing conditions for growth and development of cells for each different tissue with regard to both culture conditions and polymer design.

Cartilage

Like many other organ systems, cartilage has a limited ability to regenerate, making it an important tissue to engineer. Until recently, cartilage transplants or prosthetic devices have been used to correct cartilage defects. Prosthetic devices carry an inherent risk of infection due to the presence of a foreign body, breakdown of the interface between the prosthesis and underlying bone, and the loss of adjacent normal cartilage and bone which must be surgically removed for implantation of the device (Vacanti and Vacanti, 1991).

In 1991, our group reported the successful generation of homogeneous plates of cartilage each weighing up to 100mg using synthetic biodegradable polymer templates seeded with chondrocytes and implanted subcutaneously in nude mice (Vacanti et al, 1991). The new cartilage maintained the approximate shape and dimensions of the original templates as the polymers resorbed. Lacunae were observed in specimens harvested at 81 days and were not present in specimens harvested at 49 days or earlier, suggesting a progression from fetal to mature cartilage. Immunohistochemistry demonstrated type III collagen, usually seen in immature mammalian hyaline cartilage, in specimens from day 49 and earlier. Type II collagen, found in mammalial hyaline cartilage, was seen in specimens implanted 49 days and longer (Vacanti et al, 1991).

The next step was to engineer cartilage in specific predetermined sizes and shapes using chondrocytes seeded on synthetic polymer templates. The fabrication of tissue engineered cartilage in the form of human nasal cartilages was reported by Puelacher et al in 1993. Models of human nasal cartilage were formed using flexible, nonwoven PGA polymer mesh and exposed to a 5% solution of polylactic acid to bond the PGA fibers. The polymer forms were then seeded with a suspension of chondrocytes harvested from newborn bovine hyaline cartilage. The cell-polymer constructs were incubated seven days in vitro and then implanted subcutaneously in nude (athymic) mice for eight weeks. The specific shape and size of the implants was maintained as the polymer degraded, and the newly formed cartilaginous tissue demonstrated resilence, returning to its original configuration after bending. Histological examination of harvested specimens showed organized hyaline cartilage and the implanted cells secreted sulfated glycosaminoglycans and type II collagen (Puelacher et al, 1993).

The cartilaginous structure of the human outer ear is extremely complex and offers a significant challenge in reconstructive surgery. Prosthetic replacements composed of silicone or autologous sculpted bone grafts which are widely used today for reconstruction of the outer ear are adequate but certainly not optimal. Advances in polymer design and technology made it possible for our group to tissue engineer growth of new cartilage in the complex shape of a human ear (Vacanti et al, 1992). This work was done using polylactic acid (PLA), polyglycolic acid (PGA), and their copolymers (PLGAs). A soft nonwoven polymer mesh with 14 μm diameter fibers was formed into the shape of a human ear using a silicone prosthetic ear as a mold. A cell suspension of chondrocytes harvested from bovine hyaline cartilage was seeded onto the ear-shaped polymer forms, and after three days in vitro, each cell-polymer construct was implanted subcutaneously on the back of an athymic mouse. The cell-polymer constructs maintained the shape of a human ear throughout the seven weeks of implantation and upon harvest, revealed newly formed cartilage with resilience similar to that of a human ear (Vacanti et al, 1992). This technology of forming neo-cartilage structures which retain the precise shape and size of the three dimensional polymer scaffolds offers many applications for reconstructive surgery.

Tendon

Plastic and orthopedic surgeons are often faced with repair and reconstruction of tendon defects. Autologous tendon grafts are ideal for reconstruction; however use of autologous tendon is limited by availability and donor site morbidity. Other approaches, which include the use of homografts, heterografts, and prosthetic materials, are also associated with possible complications including infection, the risk of transmission of infectious agents by allografts, and adhesive breakdown at the host-device interface by the mechanical prostheses. The clinical need for improved treatment options has led to research to create autologous tissue engineered tendon replacements. Cao et al, 1994, reported the formation of tendon with characteristics similar to normal mature tendon using tenocyte-polymer constructs implanted in vivo. Tenocytes isolated from newborn calf shoulders using a collagenase digestion technique were seeded on sheets of polyglycolic acid polymer and implanted in athymic mice. Specimens harvested after six weeks grossly resembled normal calf tendons from which the cells had been isolated, and histological evaluation demonstrated organized collagen fibrils similar to collagen bundles of normal calf tendon. In a later experiment, tenocytes isolated from calf shoulders and knees were seeded onto strips of nonwoven PGA mesh and PGA fibers arranged in parallel with a knot at one end (Cao et al, 1995). Harvested at intervals ranging from six to twelve weeks, the rate of parallel arrangement of the tissue structures was reportedly enhanced when longitudinal polymer fibers were used. However, over time, the cells and matrix structures attained the same parallel arrangement despite the structure of the polymer used. Biomechanical tests also demonstrated a linear increase in the tensile strength of specimens over time. Eight week specimens were reported to have 30% of the average tensile strength of size-matched normal bovine tendons, and twelve week specimens showed 57% of normal tensile strength, two orders of magnitude above the starting PGA scaffold (Cao et al, 1995). This technology has potential future clinical application for reconstructive plastic and orthopedic surgery.

Bone

Experience gained with tissue engineering of cartilage was expanded in our laboratory to research in the tissue engineering of bone. Osteocytes and osteoblasts were isolated from newborn calf shoulder and seeded onto PGA polymer scaffolds. Mature bone tissue formed following implantation of the cell-polymer constructs (Mooney and Vacanti, 1993; Vacanti et al, 1993). At five weeks post implantation, the newly formed tissue resembled developing cartilage; however, vascular sources did not regress from the developing bone. The new bone became organized with evidence of trabeculae and bone plates on histologic examination. Like earlier experiments with cartilage, the new bone tissue maintained the size and shape of the polymer templates. These cell-polymer constructs offer the potential for repair of large bone defects as well as the development of vascularized bone by implantation of the constructs around a vascular pedicle.

Urothelial tissue

Shortage of native tissue has placed significant constraints on surgical reconstruction in a variety of urological conditions. A variety of natural tissues such as omentum as well as synthetic materials have been tried. Bowel has been used for urinary tract reconstruction for many years for lack of a better substitute; however this too has many associated complications including metabolic abnormalities, perforation, diverticular formation, and malignancy (Atala et al, 1992). An autologous substitute would be ideal. In 1992, Atala et al described using biodegradable polymers as delivery vehicles for rabbit urothelial cells in vivo. Newly isolated rabbit urothelial cells were noted to attach to the PGA polymers in vitro and to survive, and reorganized into multilayered epithelial structures when implanted in athymic mice, maintaining a pattern of urothelial cytokeratin expression. In 1993, Atala et al reported the development of human urothelial and bladder muscle structures by the successful harvesting and attachment of human bladder urothelial and muscle cells to biodegradable polymers. Manipulation of the cell-polymer construct into a tube before implantation demonstrated the formation of multilayered urothelial structures with a maximum thickness of 9 to 10 cell layers as in normal urothelial tissue (Atala et al, 1993).

Intestine

Children and adults who have undergone major resection of small bowel often develop a state of malabsorption and malnutrition known as "short bowel syndrome". Parenteral nutrition has improved survival; however, the use of long term hyperalimentation is associated with a number of complications including episodes of infection, loss of vascular access, and progressive liver disease. Small bowel transplantation, the most promising surgical technique, is still experimental and like transplantation of other solid organs is limited by scarcity of donor organs. Using techniques of tissue engineering developed in our laboratory, Organ et al (1993) described the development of a small animal model to create new intestinal mucosal tissue for functional replacement for absent or diseased intestine using using intestinal epithelial cells, or enterocytes, seeded on biodegradable polymer scaffolds. Following seeding on the PGA polymer, the enterocyte-polymer constructs were rolled around silastic stents and implanted on the omentum and small bowel mesentery of host animals.

Histologic evaluation demonstrated that enterocytes attached to the polymer. Following implantation, the enterocyte-polymer constructs became vascularized and remained viable over a one month period. Specimens from 14–28 days after implantation showed enterocyte proliferation and continued development of a stratified epithelium five to seven cells in thickness (Organ et al, 1993). Research in tissue engineering neointestine is continuing in our laboratory.

Heart Valves

Valvular heart disease causes significant morbidity and mortality in the United States each year, and current valve replacements using bioprosthetic or mechanical valves have significant limitations. These limitations include an inability to grow and repair, susceptibility to infection, thromboembolic complications, and limited durability. Our laboratory has recently demonstrated the feasibility of tissue engineering heart valve leaflets using autologous vascular cells and biodegradable polymers. Breuer et al, 1996, reported the development of tissue that functioned as a pulmonary valve leaflet under physiologic conditions in a lamb model. Color Doppler interrogation of the implanted leaflet showed that the leaflet functioned with only trivial regurgitation. Histology of the valve leaflet in vivo eleven weeks demonstrated that the leaflets had developed appropriate cellular architecture and extracellular matrix similar to native leaflet. The autologous leaflets maintained their size and shape after implantation in the pulmonary artery position of lamb models, and the maximal tensile strength as measured using an Instron tensometer showed values similar to native valve tissue (Shinoka et al, 1995).

The formation of thick structures

While the creation of many different tissues has been demonstrated using principles of tissue engineering, the formation of thick structures with an intrinsic vascular network has been impossible. A new technology in polymer design now exists to allow creation of thick, complex, three-dimensional tissues with the ultimate goal of creating prevascularized composite structures that can be placed like a transplanted organ or vascularized free flap. Research is being carried out in our laboratory at this time utilizing this technology.

Three-dimensional printing technology used to create three-dimensional biodegradable scaffolds by free form fabrication on copolymers of polylactide-co-glycolides using computer-assisted-design-computer-assisted-manufacture (CAD-CAM) representation of the object to drive the fabrication process. "Using these computer assisted design methods, researchers will be able to shape the plastics into intricate scaffolding beds that mimic the structure of specific tissues and even organs" (Langer and Vacanti, 1995). The use of CAD-CAM technology has the potential to be used in the future to custom fit complex devices to individual patients (Mooney and Langer, 1995). Future studies will investigate cell-polymer constructs in bioreactors under static and flow conditions and in vivo.

CONCLUSION

The year 2000 marks the dawn of a new millennium, and as we move into this new millennium, we will move beyond the realm of transplantation and into the realm of fabrication. With advances in medicine and science, we are rapidly moving toward the day when we will make whole functioning organ structures and implant them as a transplant (Langer and Vacanti, 1995).

Much new knowledge has been gained in tissue engineering across a broad front of tissue types, and there is increasing optimism that this will achieve clinical application. Today, researchers are able to construct artificial tissues that look and function like their native counterparts. In the future, using computer aided design and manufacturing methods, researchers will shape the plastics into intricate structures mimicking specific tissues and organs (Langer and Vacanti, 1995). Tissue engineering holds the potential to make tremendous impacts in medicine and reconstructive surgery.

References

Atala A, Vacanti JP, Peters CA, Mandell J, Retik AB, Freeman MR (1992): Formation of urothelial structures in vivo from dissociated cells attached to biodegradable polymer scaffolds in vitro. *J Urol* 148:658–662

Atala A, Freeman MR, Vacanti JP, Shepard J, Retik AB (1993): Implantation in vivo and retrieval of artificial structures consisting of rabbit and human urothelium and human bladder muscle. *J Urol* 150:608–612

Ben-Ze'ev A, Farmer SR, Pennman S (1980): Protein syolhern, requires cell-surface contact while nuclear events respond to cell chape in anchorage-dependent fibro-blast. *Cell* 2:365–372

Breuer CK, Shin'oka T, Tanel RE, Zund G, Mooney DJ, Ma PX, Miura T, Colan S, Langer R, Mayer JE, Vacanti JP (1996): Tissue engineering lamb heart valve leaflets. *Biotechnology and Bioengineering* 50

Burke JF, Yannas IV, Quinby WC Jr, Bondoc CC, Jung WK (1981): Successful use of a physiologically acceptable artificial skin in the treatment of an extensive burn injury. *Ann Surg* 194:413–428

Cima LG, Langer R, Vacanti JP (1991a): Polymers for tissue and organ culture. *J Bioactive Compat Polymers* 6:232–240

Cima LG, Vacanti JP, Vacanti C, Ingber D, Mooney D, Langer R (1991b): Tissue engineering by cell transplantation using degradable polymer substrates. *J Biomech Eng* 113:143–151

Cao Y, Vacanti JP, Ma PX, Paige KT, Upton J, Chowanski Z, Schloo B, Langer R, Vacanti CA (1994): Generation of neo-tendon using synthetic polymers seeded with tenocytes. *Transplan Proc* 26:3390–3392

Cao Y, Vacanti JP, Ma PX, Ibarra C, Paige KT, Upton J, Langer R, Vacanti CA (1995): Tissue engineering of tendon. *Mat Res Soc Symp Proc* 394:83–89

Folkman J, Haudenschild C (1980): Angiogenesis in vitro. *Nature* 288:551–556

Folkman J, Hochberg MM (1973): Self-regulation of growth in three dimensions. *J Exp Med* 138:745–753

Folkman J, Moscona A (1978): Role of cell shape in growth control. *Nulurc* 273:345–349

Langer R, Vacanti JP (1993): Tissue engineering. *Science* 260:920–926

Langer R, Vacanti JP (1995): Artificial organs. *Scientific American.* 273:100–103

MacDonald N (1983): *Trees and Networks in Biological Models.* New York: John Wiley and Sons

Mooney DJ, Langer RS (1995): Engineering biomaterials for tissue engineering: the 10–100 micron size scale. In: *The Biomedical Engineering Handbook,* Bronzino JD, ed. Boca Raton, FL: CRC Press

Mooney DJ, Vacanti JP (1993): Tissue engineering using cells and synthetic polymers. *Transplant Rev* 7:153–162

Mooney DJ, Cima L, Langer R, Johnson L, Hansen LK, Ingber DE, Vacanti JP (1992): Principles of tissue engineering and reconstruction using polymer-cell constructs. *Mat Res Soc Symp Proc* 252:345–352

Morse MA, Vacanti JP (1988): Time lapse photography: A novel tool to study normal morphogenesis and the diseased hepatobiliary system in children. *J Pediatr Surg* 23:69–72

Murray JE, Merrill JP, Harrison JH (1955): Renal homotransplantation in identical twins. *Surg Forum* 6:432–436

Nerem RM (1991) *Ann Biomed. Eng* 19:529

Organ GM, Mooney DJ, Hansen LK, Schloo B, Vacanti JP (1993): Enterocyte transplantation using cell-polymer devices to create intestinal epithelial-lined tubes. *Transplant Proc* 25:998–1001

Puelacher WC, Vacanti JP, Kim SW, Upton J, Vacanti CA (1993): Fabrication of nasal implants using human shape-specific polymer scaffolds seeded with chondrocytes. *Am Coll Surgeons Surg Forum* 44:678–680

Russell PS (1985): Selective transplantation. *Ann Surg* 201:255–262

Shinoka T, Breuer CK, Tanel RE, Zund G, Miura T, Ma PX, Langer R, Vacanti JP, Mayer JE Jr (1995): Tissue engineering heart valves: Valve leaflet replacement study in a lamb model. *Ann Thorac Surg* 60:S513–516

Skalak R, Fox CF (eds) (1988): *Tissue Engineering.* Liss, NY

Vacanti CA, Vacanti JP (1991): Functional organ replacement: The new technology of tissue engineering. In: *Surgical Technology,* Tawes RL, ed. London: Century Press

Vacanti CA, Langer R, Schloo B, Vacanti JP (1991): Synthetic polymers seeded with chondrocytes provide a template for new cartilage formation. *Plastic Reconstruct Surg* 88:753–759

Vacanti CA, Cima LG, Ratkowski D, Upton J, Vacanti JP (1992): Tissue engineered growth of new cartilage in the shape of a human ear using synthetic polymers seeded with chondrocytes. *Mat Res Soc Symp Proc* 252:367–374

Vacanti CA, Kim W, Upton J, Vacanti MP, Mooney D, Schloo B, Vacanti JP, (1993): Tissue-engineered growth of bone and cartilage. *Transplantation Proceedings* 25:1019–1021

Vacanti JP (1988): Beyond transplantation. *Arch Surg* 123:545–549

Vacanti JP (1993): Commentary. *Cell Transplant* 2:409–410

Vacanti JP, Vacanti CA (1996): The challenge of tissue engineering. In: *Textbook of Tissue Engineering,* Lanza R, Langer R, Chick W, eds. RG Landes

Vacanti JP, Morse MA, Saltzman WM, Domb AJ, Perez-Atayde A, Langer R (1988): Selective cell transplantation using bioabsorbable artificial polymers as matrices. *J Pediatr Surg* 23:3–9

UNOS Update (1996): 12(1):25. Richmond, VA: National Organ Procurement and Transplantation Network

2

LEGAL ISSUES INVOLVED IN TISSUE ENGINEERING: INTELLECTUAL PROPERTY ISSUES

PATREA L. PABST

Tissue engineering involves a number of complex legal issues. Those that will be addressed in this chapter are limited to intellectual property issues and interactions with the Food and Drug Administration in the United States, with a brief discussion of corresponding intellectual property issues outside of the United States.

INTRODUCTION

Intellectual property broadly includes patents, trademarks, copyrights, and trade secrets. A patent is a grant by an individual government entity to exclude others from making, using, selling, or importing into a geographical area, such as the United States, a composition or method of manufacture or use defined by the claims in the patent. A patent is awarded for a defined period of time—in most cases that time runs 20 years from the date an application for patent is initially filed. Although, in some cases, shorter terms due to disclaimers of patent term in view of earlier issued patents or longer terms due to delays relating to appeals or regulatory approval may be obtained. Trademarks are typically associated with the sale of goods or services and are used to denote the origin of the goods or services. Advantages of trademarks are that they are not limited in term and rights arise upon use. One very well-known trademark is Coca-Cola™, which has been in continuous use since before the turn of the century. The company has used that trademark in combination with retaining the formula as a trade secret

Synthetic Biodegradable Polymer Scaffolds
Anthony Atala and David Mooney, Editors
Robert Langer & Joseph P. Vacanti, Associate Editors
© 1997 Birkhäuser Boston

to create enormous value for the company. Trade secrets are also unlimited in term but must be actively protected. Trade secrets are lost if another party independently derives the same method or composition that is being maintained as a trade secret. Copyright is typically used to protect writings, although protection may extend to visual depictions of products or to advertising material associated with the use and sale of products. Additionally, copyright may protect protocols, instructions, or software programs. These intellectual property forms, in particular patents, and their application to tissue engineering are discussed in greater detail below.

PATENTS

Patents have basically the same requirements throughout the world. In the United States, the requirements for obtaining and asserting a patent are defined by chapter 35 of the United States Code ("U.S.C.").

Patentable Subject Matter

The first requirement of patentability is that the subject matter must be patentable, as defined by 35 U.S.C. § 101. In general, patentable subject matter includes compositions, methods of manufacture, and methods of use. Compositions may include dissociated cells, matrices for growing or implantation of cells, apparatus for culturing, isolation, or purification of cells, or methods and apparatus for determining various biochemical criteria of viability and function. Methods of use can entail methods for surgical implantation, for example, the creation of new tissue that can act as a bulking agent or provide a particular function. That function may be structural, as in the case of a bone implant or heart valves, or it may be biochemical, as in the case of implanted isolated cells that secrete insulin. Although the law provides for patenting of compositions, methods of manufacture, and methods of use, tissue engineering can present a problem under section 101 when the subject matter moves away from the realm of the artificial or "things engineered by the hand of man" to a blend or chimera of "artificial" and "natural". *See Diamond v. Chakrabarty,* 447 U.S. 303, 65 L. Ed. 2d 144, 100 S. Ct. 2204, 206 U.S.P.Q. 193 (1980). An example is when one blends cells and a matrix to form a cell matrix structure that is then implanted in a patient. Then, the matrix degrades to leave only implanted cells and/or the patient's own tissue grows into an implanted matrix structure which then degrades. At what point do these materials become patient and not patentable subject matter? Ethical issues may arise due to overlap between patient material and traditional subject matter, particularly in those cases involving dissociated isolated cells, biodegradable matrices for implantation, polymeric materials for altering cell/cell interaction (such as adhesion or restenosis), as well as materials for implantation that are designed to remain in the body, such as stainless steel hip replacements or cryopreserved pig valves.

Outside of the United States, methods of treatment of humans or other animals are generally not patentable subject matter. For example, although surgical instruments, drugs, or devices used in surgery are patentable, surgical treatments are not patentable subject matter. Therefore, one cannot obtain a patent on a method for surgically treating a patient. This raises the issue of whether a method for implantation of dissociated cells or cells in a polymeric matrix or other forms of tissue engineering would be patentable subject matter outside of the United States. Typically, while this subject matter is not patentable, the compositions and methods of manufacture for use in treating patients are patentable subject matter. Claims may be obtained to the composition *per se,* the dissociated cells or polymer, or the polymer matrix seeded with cells, which is to be implanted. In Europe, claims can be obtained to a first, or even a second, use of the matrix when the matrix itself is known. However, the patentability is quite limited in individual countries and in the European Patent Office for policy and ethical reasons. Generally, patent offices in asian countries are far less flexible than the European Patent Office in this matter. As a result, patent attorneys have adopted a number of strategic approaches to obtain protection equivalent to that which is available in the United States. For example, one may draft claims directed to methods of manufacture of such tissue engineering materials, as well as to methods of use that are defined by the composition rather than the method of use steps.

Novelty

The second requirement for patentability is novelty. Novelty, in its simplest terms, means that no one, including the applicant for the patent, has publicly used or described that which is being claimed, prior to filing an application for patent. In the United States, there is an exception when the publication is made less than one year prior to filing of the patent application. The publication can be "removed" as prior art if the applicants are able to demonstrate that, prior to the publication, they conceived and diligently reduced to practice what they are claiming.

What constitutes a publication? Generally, a publication is any oral, written, or physical description that conveys to the public that which applicant would like to claim. It may be a talk at the proceedings of a society (including any slides presented), an article in a scientific journal, a grant application that is awarded, a thesis, or even an offer for sale or a press release. A critical requirement is that the publication must be enabling, that is, it must convey to one of ordinary skill in the art how to make and use that which is being claimed. Public use means more than using the composition or method in one's laboratory. However, it can include even a one patient study that is reported during clinical rounds or at a presentation at which a drug company or surgical supply representative is present. The courts, in many cases, have had to interpret what it means to be publicly available. A frequent question is when is a student's thesis available as prior art. Courts have now held that once the thesis is cataloged, it is publicly available, because it has been entered into a computer database that one searching the database will be able to access. *See Philips Elec. & Pharmaceutical Indus. Corp. v. Thermal & Elec. Indus., Inc.,* 450 F.2d 1164, 1169–72, 171 U.S.P.Q. 641 (3rd Cir. 1971); *Gulliksen v. Halberg,* 75 U.S.P.Q. 252 (Pat. Off. Bd. Int'f. 1937). Accordingly, the publication date of a thesis is the date on which the thesis is cataloged, not the date on which it is defended or signed by the thesis committee. Slides that are not distributed, but that are shown at an oral presentation, are considered to be publications, particularly if the meeting is attended by those skilled in the art who would be able to understand and use the information in the slides.

Disclosures to another party under the terms of a confidentiality agreement are not publications. Uses that are strictly experimental may not be public disclosures, if, among other aspects, they are designed to determine if that which is to be claimed will work, and if any other parties who are involved are clearly informed that the studies are experimental in nature. If something is an announcement that does not enable one of ordinary skill in the art to use or make that which is claimed, then the disclosure is not a publication. For example, an announcement could be a statement made to the press that researchers X and Y have discovered a cure for cancer. Since the announcement does not tell one of ordinary skill in the art how to cure cancer, it is not enabling. However, enablement can be difficult to prove and standards may change over time. A recent court case in which the question of whether a publication was enabling related to the development of a transdermal patch for delivery of nicotine. *Ciba-Geigy Corp. v. Alza Corp.,* 864 F. Supp. 429, 33 U.S.P.Q.2d 1018 (D.N.J. 1994). The court found that a prior publication referring to transdermal patches for drug delivery mentioned that the drug in the transdermal patch for treatment of heart disease could be replaced with nicotine for assisting patients in quitting smoking. The court held that the article disclosed or made obvious the transdermal patch for delivery of nicotine claimed by the applicant, because the applicant merely took the transdermal patch described in the article, put nicotine in it, and demonstrated that the nicotine was delivered and would work exactly as the other drug did. Even though there was no information relating to the exact dosage or schedule, or how the drug was to be encapsulated, the publication was enabling because those of ordinary skill in the art would have been able to determine the dosage and how to put the nicotine in the transdermal patch without undue experimentation.

Nonobviousness or Inventive Step

The third requirement for patentability is that the claimed method or composition must be nonobvious to those of ordinary skill in the art from what is publicly known. This is usually referred to outside of the United States as a requirement for an inventive step. In the 1960s, the United States Supreme Court carefully analyzed nonobviousness and those factors that are to be considered in determining whether that which is claimed is obvious from the prior art. *Graham v. John Deere Co.,* 383 U.S. 1, 86 S. Ct. 684, 15 L. Ed. 2d 545, 148 U.S.P.Q. 459 (1966). This analysis is a fact-based determination, involving not only the elements that are claimed but also the level of skill in the art and the expectation or predictability that the claimed method or composition would perform as predicted, its actual success in the marketplace, long felt need, and whether there are unexpected results. If one has no better than a 50-50 chance that a particular method may work and the method works, it is arguably not obvious, although it may be obvious to try. If one tries something and the results are vastly different from what was expected, then the results are not obvious. For example, if one administered two drugs each in the dosage known to yield a particular effect and the combination resulted in a substantially greater effect than the sum of the two drugs, providing the ability to use a much lower dosage of each drug than expected, then one would have unexpected results or "synergy". If the prior art teaches away from what the applicant has done, then this result would support a finding of nonobviousness. For example, if the prior art says that one cannot administer drugs transdermally using ultrasound except at a very high frequency, then it may be nonobvious if the applicant for a patent finds that the same or better results are obtained using a very low frequency. Many other considerations factor into whether a claimed composition or method is obvious in view of the prior art.

Enablement and Best Mode

The fourth requirement for patentability is enablement and disclosure of the best mode. These are defined by 35 U.S.C. §112. A written description is also required, but enablement and the written description requirement are usually considered together. In order to obtain a patent, the applicant must describe that which is claimed in sufficient detail, and with appropriate methods and sources of reagents or other materials or equipment, to enable one of ordinary skill in the art to make and use that which is claimed. This sounds far simpler than it actually is in practice. In many cases, particularly when coming out of a university study or a start-up company, the invention that applicants would like to claim is that which the applicant intends to develop over the next several years, based on a limited amount of data available at the time of filing. Particularly in the case of universities, where the applicant must publish or has submitted grant applications (which in and of themselves constitute prior art once they are awarded), the difficulty is in describing that which has not yet been done. The application must not only describe a specific limited example, but must describe the various ways in which one intends to practice that which is claimed. The purpose of a patent is to exclude the competition from making and using that which is claimed, not to "protect" a product—a frequent misconception of patents. In order to exclude competition, one must describe and claim not only that which one intends to practice but that which another party could practice in competition with the patentee. What does this mean in real terms? It means that the applicant for a patent must describe his preferred method, which is known as of the date of filing, the preferred embodiments that he or his company intends to market, as well as any embodiments that a competitor could make and use in competition with the applicant's product.

"Invention" usually consists of two steps, "conception" and "reduction to practice." There are two kinds of reduction to practice, actual and constructive. Constructive practice means that the applicant has described in the application for patent *how* to make and use that which is claimed, but has not actually made and used what is claimed. This may be as simple as saying that although a biodegradable polymer such as polylactic acid-co-glycolic acid is preferred for making a matrix for implantation, other biodegradable polymers such as polyorthoesters or polyanhydrides could also be used. It may be less obvious that other cell types may be used when only one example showing reduction or practice of a single cell type is available. The rule of thumb in this case is the level of predictability. Is it predictable that hepatocytes implanted in a matrix in which chondrocytes have been shown to survive and proliferate will also survive and proliferate? The answer in some cases will be "no." On the other hand, it may be more predictable that chondrocytes will survive and proliferate in a matrix in which hepatocytes have previously been shown to survive and proliferate. Therefore, in stating what kind of cells one could implant using the claimed technology, one might list a wide variety of cells based on the hepatocyte data that may or may not be possible to list based on the chondrocyte data. Being too predictive (i.e., engaging in extensive constructive reduction to practice), which includes "nonenabling" or nonenabled technology, may in some cases be a detriment during prosecution of subsequently filed applications because the examiner may cite the earlier work as making obvious the applicant's subsequent work. Patent attorneys must frequently play a balancing game in determining how far to go with constructive reduction to practice in order to exclude competitors while not eliminating the applicant's own ability to obtain additional, subsequent patent protection.

In the United States, there is a requirement to disclose the best mode for practicing that which is claimed at the time of filing the application. No similar requirement exists outside of the United States. Because most applicants file the same application in the United States as outside of the United States, U.S. applicants frequently disclose their best mode in foreign-filed applications. Furthermore, U.S. applications are not published until they are issued as a patent, while applications filed in other countries are published eighteen months after their earliest priority date. Thus, it may be desirable in some cases to omit the preferred embodiment as of the date of the foreign filing in order to prevent one's competitors from knowing the best mode for practicing the invention until the U.S. patent is issued, which may be many years after publication of the corresponding foreign application. However, this strategy may create a problem, as patent laws vary from country to country. In particular, Japan requires one to provide examples of that which one intends to claim, limiting available protection.

Duty of Disclosure

Another unique requirement of the United States patent law is the duty of disclosure, described by Chapter 37 of the Code of Federal Regulations ("C.F.R."). As defined by 37 C.F.R. §§ 1.56, 1.97, 1.98, applicants in the United States are required to submit to the examiner in the U.S. Patent Office copies of all publications or other materials that may be determined by an examiner to be material to examination of the claimed subject matter. Foreign publications must be accompanied by an English translation if they are not in English, although in some cases this may be limited to an abstract. Failure to cite relevant prior art to the Patent Office can result in a subsequent finding by an appropriate court of jurisdiction that the patent is invalid for fraud and violation of the duty of disclosure. Under the current standard, the applicant is not required to describe the relevance of the cited publication but may merely cite the publication and provide a copy to the examiner for the examiner's review. When many publications are being cited to the Patent Office, it may facilitate review to group the publications or even distinguish those that the applicant believes are most relevant. As discussed above, prior art includes oral or written publications made prior to applicant's filing date by applicant or any other entity. Relevance of material is more difficult to define.

Procedure

When a patent application is filed in the United States, it will be assigned to an examiner for review of the relevant prior art and for prosecution. The Examiner will likely issue one or more office actions, objecting to the specification or rejecting the claims as lacking novelty, being obvious, not enabling the breadth of the claims, being indefinite, or a combination thereof. The applicant can amend the claims and present arguments and supporting data to overcome the rejections. The specification, or description, cannot be amended to add new matter once the application is filed. In many cases, an agreement will be reached as to what claims are allowable, and a patent will issue. In the event that the Examiner finally rejects the claims and the applicant has exhausted his opportunities to respond to the rejections, an appeal to the Board of Patent Appeals and Interferences can be filed. A decision by the Board, which can be based not only on argument and data submitted during the prosecution of the application but also an oral hearing, will typically require two to three years. The applicant's only recourse following a negative decision by the Board is to file an appeal in the Court of Appeals for the Federal Circuit.

In the United States, patent applications are typically examined by a single examiner. In the European Patent Office, examination may be conducted by an examiner other than the one who conducted the initial search and issuance of a search report. If the applicant in the European Patent Office requests oral proceedings, at which oral argument may be presented, the oral proceedings are held before a panel, not just the examiner. In contrast, U.S. proceedings are between the applicant and the examiner; although, the examiner's supervisor may be asked to attend if an in-person interview is held. In all cases, there are appeal procedures available if an examiner maintains his rejection of the claims, asserting the claims lack novelty, are obvious, or are insufficiently enabled.

Inventorship

The U.S. Constitution, Art. 1, § 8, cl. 8, provides that inventors have the exclusive right to their discoveries. An application for patent must be made by the inventor, or under certain circumstances (such as when the inventor is dead) by persons on behalf of the inventor, under 35 U.S.C. § 111 (1988). When more than one person made the invention, the inventors are required to file jointly, "even though they did not physically work together or at the same time, each did not make the same type or amount of contribution, or each did not make a contribution to the subject matter of every claim of the patent." 35 U.S.C. § 116 (1988). A patent is invalid if it names one who is not an inventor or if it fails to name an inventor; however, these errors may be corrected if they were not committed with an intention to deceive. 35 U.S.C. § 256 (1988).

In a "nutshell", an inventor is one who conceives and/or reduces to practice the claimed invention, not at the direction of another. University settings are somewhat unique when it comes to determining what is prior art and what is the invention of another, because it is often difficult to determine who is the inventor. When one attempts to remove prior art by demonstrating that one has conceived and reduced practice prior to the publication, one must first determine who the inventors are and whether the publication is in fact a publication by another entity. If the publication is the inventors' own, then it is easier to swear behind because they must have conceived and reduced to practice prior to that publication. Merely because a publication is coauthored by one of the inventors does not mean that the publication is the inventors' own work. There must be complete identity between the named authors and the inventors for the publication to be the inventors' own work. Publications typically will include coauthors who do not meet the legal definition of an inventor.

To determine who the inventors are, one must first ascertain that which is claimed. Second, one must determine what is already in the prior art; one is not an inventor if the claimed matter is already in the prior art. For example, if one is claiming a cell matrix structure and the claim defines the matrix structure as formed from biodegradable polymer, then this particular element is probably already in the prior art and that description alone would not be the invention of any named inventor upon the application for patent. However, if the polymeric matrix were defined as having a particular structure or shape or composition that has not previously been defined, then the individual (or individuals) who determines that shape or structure or composition would be an inventor. In methods for manufacture, the person who is in the laboratory using the method may or may not be an inventor. If this person has been told by another to go and make composition X using steps A, B, and C, then the person who performs the method is not an inventor, even if there is some optimization of the concentration or selection of reagents or conditions under which they are combined. If, however, that person determines that it is essential to use a concentration ten times greater than what he has been told in order to make it work, then he may be an inventor of the method of use. A patent may name multiple parties as inventors. They do not all have to be inventors of each and every claim that defines the invention. One person may be an inventor of composition claims, another the method of manufacture claims, and yet another the method of use claims. Inventorship may need to be corrected following a restriction requirement or after cancellation or amendment of the claims.

The definition of inventor elicits questions about the definitions of "conception" and "reduction to practice." In a university research laboratory, has a graduate student participated in the conception of his professor's invention if he conducted the experiments under her direction that resulted in the reduction to practice, i.e., the synthesis of the claimed protein? Did the professor conceive the invention if she thought of the general idea and desired result but left it to the graduate student to figure out how to synthesize the protein? Is the professor's conception of a protein and the detailed steps of producing it an invention if she has not yet actually reduced her idea to practice by synthesizing the protein? And in a different laboratory, is a technician an inventor if he added bearings to the design that the supervising engineer told him to build? In each case, who, if anyone, is an inventor?

Conception is "the formation in the mind of the inventor of a definite and permanent idea of the complete and operative invention as it is thereafter to be applied in practice," such that a person of ordinary skill in the art would be enabled to convert the idea to tangible form without extensive research or experimentation. *Mergenthaler v. Scudder,* 1897 C.D. 724, 731 (D.C. Cir. 1897); *In re Tansel,* 253 F.2d 241, 242 (C.C.P.A. 1958). Usually, one who conceives the inventive idea, or part of it, is an inventor, while those who perform the ordinary interim experimentation under the direction of the inventor in reducing the inventive concept to practice are not defined as inventors. Conception followed by reduction to practice is a common sequence during invention, but invention may result from other sequences of events. For example, conception and reduction to practice may occur together, or unexpected discovery coupled with recognition of the discovery as something new and useful may occur without previous conception. For invention to be complete, the inventor's conception must include a clear idea of how the new product, process, or machine can be reduced to practice, or put into a tangible form. Reduction to practice may be done by persons other than the inventor if they work under his or her direction. In both university and commercial laboratories, an invention may occur through discovery without any previous conception. This type of invention is not formulated by conception of the inventive idea followed by reduction to practice, as described above. Rather, the invention occurs through discovery coupled with recognition that the product, process, machine, or combination is new and useful. Experimentation involved in reduction to practice may also be part of conception.

While coauthors may choose not to be named on a paper, the law requires that every person who contributes any part of what is claimed in a patent application must be named. If, in the course of prosecution of the patent application, all claims that reflect the contribution of one inventor are cancelled or rejected, that inventor's name must also be removed. There is an implicit lack of equality in coauthorship. Generally, the first or last listed author is considered to be the primary originator of the new ideas and data in the paper, and the others are assumed to be secondary collaborators. Joint inventors, however, are equal in the rights of patent that accrue to them, unless they agree otherwise. *See* 35 U.S.C. § 262 (1988). Even though they did not conceive exactly the same idea together, or each created a different part of the whole invention, or the contribution of one was only a small but essential part of the invention, all are joint inventors and share an equal right to exclude others from making, using, or selling the claimed invention. In practice, university policy usually dictates that joint inventors assign ownership of the patent to the university, while the royalties that may accrue if the patent is licensed and the invention is marketed are allocated between the university and the inventors. The inventors' share of the royalties is distributed equally among them, unless they have contracted otherwise.

Restriction Practice

Under 35 U.S.C. § 121, a patent must have claims directed solely to a single invention. Guidelines for determining what constitutes a single invention are published in the Manual of Patent Examining Procedures (MPEP). The Examiner may restrict the claims into more than one group, if, in his view, the claims are directed to patentably distinct inventions. For example, if the applicant has claims to a composition, a method of making that composition, and the method for using the composition, an examiner in the Patent Office of the United States will probably determine there are multiple inventions defined by the claims in the application as filed, and the examiner will require cancellation of the claims to all but the method of manufacture or use or composition from the application. The applicant then would be allowed to prosecute the elected claims in that application and file what are referred to as divisional applications with the nonelected claims, which may ultimately issue as a second or third patent. The effective priority date would still be the original filing date.

Patent Term

Under the revised United States patent law that was enacted as a result of implementation of the General Agreement for Trade and Tariffs (GATT), the term of a patent is twenty years from the original date of filing. Applicants therefore have more incentive to prosecute all claims in a single application in order to minimize costs for prosecuting and maintaining the patent. Under the law in effect prior to June 8, 1995, the patent term was seventeen years from the date of issue in the United States. Divisional applications were a commonly used method to extend patent protection to encompass different aspects of the technology over a period of time much greater than seventeen years. For example, an application would be filed in 1990, and a single inventive concept (e.g., the composition) would be prosecuted in the first application. Three years later, when those claims were allowable and a patent was to issue, a divisional application would be filed with another set of the claims that had been restricted out of the original application. This divisional application would be prosecuted for another two to three years, the claims would be determined to be allowable, the second patent would issue with a seventeen year term, and a third divisional application would be filed. The result is that patents on related technology would issue sequentially over several years, increasing the effective term of patent protection beyond twenty years. Under the new law, this mechanism to extend patent protection is not possible.

The GATT was signed into law in the United States on December 7, 1994, and the initial provisions affecting U.S. patent practice were implemented

June 8, 1995. The most significant changes arising from enactment of that agreement, now called the Uruguay Round Act, were changes in the patent term in the United States, the implementation of provisional patent applications, and the broadening of what constitutes infringement in the United States. The change in patent term has been discussed above. For those applications filed before June 8, 1995, the term of any issuing patent is seventeen years from the date of issue or twenty years from the filing date, whichever is longer. The term of any patent issued on an application filed June 8, 1995 or later is twenty years from the earliest claimed priority date. Extensions of terms are available upon delays in issuance arising from appeals or interferences. Additional extensions of terms are available for delays in obtaining regulatory approval by the Food and Drug Administration for a device or a drug.

Because tissue engineering is a complex field, especially in the areas of patentable subject matter and enablement, many of the general observations relating to patents may not be as directly applicable to more "conventional" patentable subject matter. For example, the U.S. Patent Office has consistently maintained that the change from a seventeen year term from the date of issue to a twenty year term from the earliest priority date will not result in a significant loss of patent right. However, because the Patent Office applies such a stringent examination proceeding under §112 (enablement) in the biotechnology area, its assertion is not likely to be true. Issue time in these cases typically has been considerably longer, not uncommonly taking as many as five to seven years from the original priority date. The result is that these complex biotechnology patents will have a substantially shortened term as compared to many other types of patents. Because a patent extension can still be obtained for delays due to regulatory issues involving the FDA, as well as for appeals to the Board of Patent Appeals and Interferences, those in the United States who believe that their patent rights will be limited in term due to delays in prosecution should avail themselves of the Patent Extension Act, if at all possible. One must bear in mind, however, that an extension for regulatory delays can only be obtained on *one* patent for any particular product or process; thus, the inventor or licensee with multiple, related patents clearly should choose the most important patent or the patent subject to the greatest increase in patent term when facing such a situation. The patent that is to be extended must be brought to the attention of the FDA. Following FDA approval of the claimed product or process, the extension must be applied for in a timely fashion.

Provisional Patent Applications

Provisional patent applications, while new to the United States, have been utilized for many years in other countries, such as the United Kingdom and

Australia. These applications are a mechanism for obtaining a filing date at minimal cost and with fewer requirements for completeness of the application and determination of the inventive entity for a period of one year. The provisional application ceases to exist twelve months after the date of filing. If an application is filed as a provisional application, it can be converted to a standard utility application at any time during the twelve months period after filing. In the alternative, it can serve as a basis for a claim to priority in a subsequently filed utility application, if the utility application is filed prior to the expiration of the one year life of the provisional application.

Although touted as a great benefit to the small entity or individual applicant, provisional applications have the same requirements for disclosure pursuant to 35 U.S.C. §112 as a standard utility application. Failure in the provisional application to completely disclose and enable that which is subsequently claimed in an utility application can result in a loss of the claim to priority to the provisional application, if that which is claimed is not enabled. Merely filing an article that will be published or presented in order to avoid loss of foreign rights usually will not comply with the enablement requirements, and therefore will not serve as an adequate basis for priority. It is essential that applicants who file provisional applications based on an article amplify the description to encompass other embodiments and to provide the basis by which one of ordinary skill in the art can practice that which is claimed. Application sections that are not required for enablement, which are typically included in a utility application, include the background of the invention, the problems that the claimed invention addresses, and the claims. These sections can be omitted from the provisional application, thus saving time and money in preparing the application. In many cases, fairly standard language can be used to amplify the description in an article in order to meet the enablement requirements, providing a means for those with limited amounts of time or money to protect that which they are disclosing with minimum risk and expenditure.

It has recently been confirmed by the World Intellectual Property Organization (WIPO), which implements the provisions of the Patent Cooperative Treaty (PCT) and the European Patent Office, that U.S. provisional applications serve as an adequate basis for a claimed priority in corresponding foreign file applications. However, under the Patent Convention, all foreign applications that claim priority for earlier filed application must still be filed within one year of the U.S. filing date or the filing date of the country in which the first application is originally filed.

Recordkeeping

It is extremely important in the intellectual property area that proper records be maintained. Laboratory notebooks should be filed and maintained in chronological order. Notebook pages should be consecutively numbered, entries should be recorded in ink, and each page should be signed and dated by the party making the entry. Furthermore, it is preferred that each page be witnessed at a time contemporaneous to the date the entries are made by one who is qualified to understand what is being entered on the page. The party does not have to completely understand every aspect of the entry; he or she merely must be qualified to say, with some degree of certainty, what the entries on that page were.

Laboratory notebooks are used to show the origin of an invention. This may be important if there is a dispute as to ownership, if there is a dispute as to inventorship, or to prove priority of inventorship where, in the United States, patent rights are awarded to the first party to invent as opposed to the first party to file a patent application. Proceedings conducted by the U.S. Patent Office to determine priority of inventorship are referred to as "interferences." The laws and regulations governing interferences were changed following enactment of the GATT to allow parties other than those residing and inventing in the United States to obtain a patent in a dispute with another party also claiming the same subject matter, by demonstrating that they were the first to conceive and reduce to practice. Interference law is very complicated. In the biotechnology area, the dates of inventorship are frequently within weeks of each other. Moreover, the courts have increasingly altered the law regarding what is conception and reduction to practice as applied to the biotechnology field, following decisions in cases such as *Amgen, Inc. v. Chugai Pharmaceutical Co.,* 927 F.2d 1200 (Fed. Cir. 1991); *Fiers v. Revel,* 984 F.2d 1164 (Fed. Cir. 1993); and *Colbert v. Lofdahl,* 21 U.S.P.Q.2d 1068 (Bd. Pat. App. & Int'f. 1995). As a result, it has become much more difficult to predict what acts will ultimately be critical in showing priority of inventorship. Accordingly, accurate record keeping by all members of a research group is extremely important.

Infringement

In addition to changes in patent term and creation of provisional patent applications, passage of the GATT changed the definition of infringement in the United States. As defined by 35 U.S.C. § 271, one who, without authority, makes, uses, offers to sell, or sells any patented invention, within the United States or imports into the United States any patented invention during the term of the patent therefor, infringes the patent. In the United States, a claim for infringement cannot be made until after issuance of the patent. In some other countries, including the European Patent Convention countries, translated claims can be filed prior to issuance of the patent and damages can be backdated to the date of filing the translated claim, once the patent issues. In the United States, a patent application is secret until it is issued as a patent, at which point it is published and can be asserted against parties whom the patent owner or exclusive licensee of the patent believes are infringing the claims, that is, against those they believe are making, using, selling or importing subject matter falling within the scope of the issued patent's claims.

A party who believes that an issued U.S. patent is not valid may file a request for reexamination, citing art that was not made of record during the prosecution of the patent. If the patent is asserted against the party, that party may go into federal district court and ask for a declaratory judgment that the patent claims are invalid or that they are not infringed. In Europe, and in many other countries, there is a post-grant opposition proceeding available. In the European Patent Office, there is also a process whereby one may file observations during the prosecution of an application, which is public, unlike in the United States. Third party observations can be used as a means to bring relevant prior art, mischaracterized prior art, or problems relating to enablement, to the attention of the European patent examiner.

Patents give the patentee, or its licensee, the right to exclude others from making, using, selling, or importing a product or process defined by the claims of an issued patent, within the country issuing the patent. In the United States, these rights are defined by 35 U.S.C. §271. 35 U.S.C. §287, which defines the limitations on damages and other remedies, was amended on September 30, 1996, in an apparent effort to meet the intended goal of the bill last reported here as pending in the U.S. legislature to prevent issuance of patents directed to new medical treatments. The amendment[1], which applies only to patents issued after September 29, 1996, deprives patentees of remedies for direct infringement and induced infringement of patents to surgical or medical procedures that do not involve patented drugs or devices. Although the exact scope of the exclusion is subject to interpre-

---

1.   Section 287 of 35 U.S.C. was amended by adding at the end the following new subsection:

(c)(1) With respect to a medical practitioner's performance, of a medical activity that constitutes an infringement under section 271 (a) or (b) of this title, the provisions of sections 281, 283, 284, and 285 of this title shall not apply against the medical practitioner or against a related health care entity with respect to such medical activity.

(2) For the purposes of this subsection:

(A) the term "medical activity" means the performance of a medical or surgical procedure on a body, but shall not include (i) the use of a patented machine, manufacture, or composition of matter in violation of such patent, (ii) the practice of a patented use of a composition of matter in violation of a biotechnology patent.

(B) the term "medical practitioner" means any natural person who is licensed by a State to provide the medical activity described in subsection (c) (1) or who is acting under the direction of such person in the performance of the medical activity.

(C) the term "related health care entity" shall mean an entity with which a medical practitioner has a professional affiliation under which the medical practitioner performs the medical activity, including but not limited to nursing home, hospital, university, medical school, health maintenance organization, group medical practice, or a medical clinic.

tation of numerous terms in the amendment, it appears that the exclusion applies only to the performance by a medical practitioner (defined as a person licensed to provide medical activity or a person under the direction of such a person) of a medical or surgical procedure not involving the use of a patented machine, manufacture, or composition of matter (e.g. medical devices, implants, and drugs) or the patented use of a composition of matter. Thus it appears that the use of patented medical devices and patented drugs will still be subject to infringement remedies.

There is an exception included in the amendment whereby the performance of some patented uses of compositions of matter (e.g. patented uses of implants and drugs) cannot form the basis for seeking remedies for infringement. These are methods "where the use of the composition of matter does not directly contribute to the achievement of the objective of the claimed method." While it is not clear how this exception will be interpreted, it presumably prevents remedies for infringement in the case of, for example, a method of resectioning an organ where the use of an antiseptic is recited. In this example, although the antiseptic is a "composition of matter" in most case it would not be considered to "directly contribute to the achievement of the objective of the claimed method. "

It should be noted that a patentee is prevented from seeking remedies for

continued from page 29

(D) the term "professional affiliation" shall mean staff privileges, medical staff membership, employment or contractual relationship, partnership or ownership interest, academic appointment, or other affiliation under which a medical practitioner provides the medical activity on behalf of, or in association with, the health care entity.

(E) the term "body" shall mean a human body, organ or cadaver, or a nonhuman animal used in medical research or instruction directly relating to the treatment of humans.

(F) the term "patented use of a composition of matter" does not include a claim for a method of performing a medical or surgical procedure on a body that recites the use of a composition of matter where the use of that composition of matter does not directly contribute to achievement of the objective of the claimed method.

(G) the term "State" shall mean any state or territory of the United States, the District of Columbia, and the Commonwealth of Puerto Rico.

(3) This subsection does not apply to the activities of any person, or employee or agent of such person (regardless of whether such person is a tax exempt organization under section 501(c) of the Internal Revenue Code), who is engaged in the commercial development, manufacture, sale, importation, or distribution of a machine, manufacture, or composition of matter or the provision of pharmacy or clinical laboratory services (other than clinical laboratory services provided in a physician's office) where such activities are:

(A) directly related to the commercial development, manufacture, sale, importation, or distribution of a machine, manufacture, or composition of matter or the provision of pharmacy or clinical laboratory services (other than clinical laboratory services provided in a physician's office), and

(B) regulated under the Federal Food, Drug, and Cosmetic Act, the Public Health Service Act, or the Clinical Laboratories Improvement Act.

(4) This subsection shall not apply to any patent issued before the date of enactment of this subsection.

infringement against both the medical practitioner and any related health-care entity (defined as an entity with which the medical practitioner has a professional affiliation) for performance by the medical practitioner of the medical or surgical procedures recited in the amendment.

The amendment does not apply to activities of persons engaged in the commercial development, manufacture, sale, importation, or distribution of a machine, or composition of matter or the provision of pharmacy or clinical laboratory services where such activities are (1) directly related to the commercial development, manufacture, sale, importation, or distribution of a machine, manufacture, or composition of matter or the provision of phar-macy or clinical laboratory services, and (2) regulated under the Federal Food, Drug, and Cosmetic Act, the Public Health Service Act, or the Clinical Laboratories Improvement Act. Although it is not clear how this exclusion will be interpreted, it appears that infringement remedies will not be limited for performance of a medical or surgical procedure that is related to the commercial development or exploitation of a product or service where the medical or surgical procedure is subject to Federal regulation.

From a practical perspective, patents are usually only enforced against manufacturers or distributors, not doctors, since it would be unduly burden-some, and typically not cost effective, to pursue actions against many indi-viduals rather than a single large entity. It is therefore doubtful the change in law will have any detrimental impact on the health care industry in the United States. It is certainly relevant in the area of tissue engineering, however, and its effect may yet be felt in this rapidly evolving area.

TRADEMARKS, TRADE SECRETS, AND COPYRIGHTS

Other types of intellectual property that may have applicability to tissue engineering include trade secrets, copyrights, and, to a lesser degree, trade-marks. Trademarks will not be discussed extensively here, other than to again note that rights in trademarks arise from use and are used to designate the origin of the goods. A company name as well as a product name can be a trademark. The trademark cannot be generic or totally descriptive of the product, and it must be distinct enough from other trademarks in a similar field of use or similar good or service to avoid any likelihood of confusion as to the origin of the good or service among the consumers of the trade-mark good or service.

Trade secrets can be compositions or methods of manufacture or even uses that are maintained in secrecy. Most companies that have optimized methods for manufacture (e.g., methods for culturing especially difficult cells or for processing polymers to impart the most desirable physical and chemical properties) keep them secret. In order to maintain the process or product as a secret, one must (1) not disclose the process or product in public, and (2) must take affirmative steps to protect the information from public disclosure. This duty includes informing parties who may accidently become aware of the technology, as well as those who are intentionally informed regarding the technology, that the material is a trade secret and is to be maintained in confidence. Laboratory notebooks describing processes or products that are considered proprietary should be maintained in designated areas labeled confidential or restricted access. Employees involved in the use of the trade secrets should be informed that the material is to be maintained as confidential and that breach of any agreement with the company by disclosing the trade secrets to a third party could result in irreparable harm and therefore be subject to injunctive relief. Trade secrets cease to be trade secrets upon public disclosure, as discussed above, or when they are independently developed by another party. If a third party obtains the trade secret, the original holder of the trade secret has no recourse unless he can prove that the secret was acquired by theft, fraud, or other improper means. Unlike patents, which have a defined term during which the patentee can exclude others from competition, trade secrets are subject to no similar limitation.

Copyrights are used to protect written or visual materials. These can include computer software programs, publications, protocols, or other materials. In many cases where the author is employed or engaged as a consultant, the copyrights will be owned by the party contracting with the author, the journal publishing the work, or the employer. Copyrights can be used to prevent exact copying; they do not protect an "idea". They are transferrable and enforceable under U.S. law and in many foreign jurisdictions, and they can be extremely valuable.

CONCLUSIONS

Intellectual property rights provide a means for the owners of technology to recover their investment in the technology and, in some cases, to make a profit. More importantly, the intellectual property rights provide a means for financing the incredibly expensive research and development and testing required for commercialization of new products and processes in the medical and biotechnology field. When the intellectual property rights have been lost or given away by publication, many times it is not possible to obtain the money required to see a product or process reach the clinic and benefit those for whom it is intended. Despite those who decry the moral questions of why entities should profit from the misfortune of others through the patenting and other forms of protection of the technology, it is only by protecting the technology that it can be used to help those who need it the most.

3

LEGAL ISSUES INVOLVED IN TISSUE ENGINEERING: FDA REGULATION OF TISSUE ENGINEERING

WILLIAM H. KITCHENS

INTRODUCTION

Emerging therapies involving cell and genetic materials and their commercial development challenge the limits of the U.S. Food and Drug Administration's (FDA) regulation of these products as drugs, medical devices, and biologic products. Existing FDA statutory authorities, although enacted prior to the advent of tissue engineering, have been used by FDA to encompass a regulatory scheme for these new products and to require that areas such as safety, efficacy, quality control, and potency be thoroughly addressed prior to marketing. In a rapidly evolving field like cell and gene therapy, regulatory policies and approaches can be expected to change as new products are developed and experience accumulates; however, some general observations can be made about the agency's current regulatory strategies in this area.

GENERAL FRAMEWORK FOR REGULATION OF BIOLOGICS, DRUGS, AND DEVICES

FDA regulates numerous kinds of products intended to prevent, treat, or diagnose diseases or injuries under a legal scheme established in the Federal Food, Drug, and Cosmetic Act (FDCA) and the Public Health Service (PHS) Act. These existing statutory provisions have been applied to a vari-

Synthetic Biodegradable Polymer Scaffolds
Anthony Atala and David Mooney, Editors
Robert Langer & Joseph P. Vacanti, Associate Editors
© 1997 Birkhäuser Boston

ety of products developed by tissue engineering. FDA has regulated these products as "biologics," "drugs," or "devices."

Section 351(a) of the PHS Act (42 U.S.C. § 262(a)) identifies a biological product as "any virus, therapeutic serum, toxin, antitoxin, vaccine, blood component or derivative, allergenic product, or analogous product, or arsphenamine or its derivatives (or any other trivalent organic arsenic compound), applicable to the prevention, treatment, or cure of diseases or injuries of man." Products considered to be "biological products" under the PHS Act are also considered "drugs" or "devices" subject to the applicable provisions of the FDCA. For example, the adulteration, misbranding, and registration provisions of the FDCA would also apply to the biological product as a drug or device.

Section 201(g)(1) of the FDCA (21 U.S.C. § 321(g)(1)) defines the term "drug," to include "articles intended for use in the diagnosis, cure, mitigation, treatment, or prevention of disease in man or other animals." Similarly, the term "device" is defined in the FDCA as "an instrument, apparatus, implement, machine, contrivance, implant, in vitro reagent, or other similar or related article . . . intended for use in the diagnosis of disease or other conditions, or in the cure, mitigation, treatment, or prevention of disease in man or other animals . . . which does not achieve its primary intended purpose through chemical action within or on the body of man or other animals and which is not dependent upon being metabolized for the achievement of its primary intended purpose." Both the "drug" definition and the "device" definition under the FDCA also include articles "intended to affect the structure or any function of the body."

Under Section 501 of the FDCA (21 U.S.C. § 351), both drugs and devices are considered adulterated for a number of specified reasons, including that the product's strength or purity fall below established standards, the product consists in whole or in part of any filthy, putrid, or decomposed substance, the container for the product is composed of a poisonous or deleterious substance that may render the contents injurious to health, or the product has been prepared, packed, or held under insanitary conditions. Moreover, a product will be considered adulterated if the methods and facilities and controls used for its manufacture, processing, packing, or holding fail to conform with current good manufacturing practice (CGMP) regulations. FDA's implementing regulations codified at 21 C.F.R. Parts 211 and 820 specify the drug and device CGMP requirements.

The misbranding provisions of the FDCA that apply to drugs and devices are set forth in Section 502 of the FDCA (21 U.S.C. § 352). Under this statutory authority, a drug or device is considered misbranded for a number of specified reasons, including labeling that is false or misleading in any particular or labeling that fails to bear adequate directions for use or adequate warnings against unsafe use (21 U.S.C. § 352(a) and (f)). A drug or device is also misbranded if it is dangerous to health when used in the manner or with the frequency suggested in the labeling (21 U.S.C. § 352(j)). For prescription drugs and restricted devices, certain information must be included in all advertisings or other printed material (21 U.S.C. § 352(n) and (r)). Labeling and advertising requirements are set forth in great detail in FDA regulations. (21 C.F.R. Parts 201, 202, and 801).

Section 510 of the FDCA (21 U.S.C. § 360) requires persons who own or operate establishments for the manufacture, preparation, propagation, compounding, or processing of drugs or devices (with certain exceptions) to register those establishments with FDA. Individuals who must register their establishments must also file a list of all the drugs and devices being made or processed at the establishment. FDA's registration regulations are codified at 21 C.F.R. Parts 207 and 807.

PREMARKET APPROVAL REQUIREMENTS

New drugs

Products in the tissue engineering field that are regulated as "drugs" by FDA must be approved through the new drug application (NDA) process as safe and effective. As a practical matter, the role of FDA in the early stages of development of drug products is limited. In general terms, the FDA's first involvement is to determine whether the drug is safe enough to test in humans, and, if so, after all human testing is complete, to decide whether the drug can be sold to the public and what its label should say about directions for use, side effects, warnings, and the like. FDA's involvement with the drug sponsor begins after testing on animals. Although FDA usually does not mandate what specific laboratory or animal studies are required, the agency does have regulations and guidelines on the kinds of results FDA expects to see before the sponsor seeks permission to conduct human testing.

The actual approval process by FDA begins with an investigational new drug (IND) application, filed before the first human experimentation with the drug. The IND process allows FDA an opportunity to examine the proposed clinical studies for the drug. FDA regulations establish the contents and format for INDs (21 C.F.R. Part 312). An IND may be submitted for one or more phases of clinical investigation. The clinical investigation of a previously untested drug is generally divided into three phases. Although, in general, the phases are conducted sequentially, they may overlap. Phase I includes the initial introduction of an investigational new drug into humans. These studies are typically closely monitored and may be conducted in patients or normal volunteer subjects. Phase I studies are designed to determine the metabolism and pharmacologic actions of the drugs in humans, the side effects associated with increasing doses, and, if possible, Phase I studies seek early evidence on effectiveness. The total number of subjects and patients included in Phase I studies varies with the drug but is generally in the range of 20 to 80.

Phase II studies are conducted to evaluate the effectiveness of the drug for a particular indication or indication in patients with the disease or condition under study and to determine the common short-term side effects and risks associated with the drug. Phase II studies are typically well-controlled, closely monitored, and conducted in a relatively small number of patients, usually involving no more than several hundred subjects.

Phase III studies are expanded controlled and uncontrolled trials. They are performed after preliminary evidence suggesting effectiveness of the drug has been obtained and are intended to gather the additional information about effectiveness and safety that is needed to evaluate the overall benefit-relationship of the drug and to provide an adequate basis for physician labeling. Phase III studies usually include from several hundred to several thousand patients.

If the IND application is not acted upon in thirty days, either by approval, disapproval, or an FDA request for a delay in the administration of the drug to the patients, then the IND is deemed accepted and the clinical studies may begin. FDA controls over human clinical investigations are supplemented with regulations on clinical research practices, responsibilities of sponsors and investigators, institutional review boards (IRBs), special protections for certain classes of test subjects, and other similar measures.

Following completion of the clinical studies, a new drug application (NDA) must be prepared and submitted to FDA. Detailed regulations establish both the contents and format for NDAs and the procedures to be followed by the drug sponsor. FDA regulations provide that within 180 days of receipt of the NDA, the agency must act on the application, although as a practical matter the review process is normally longer. Improvement in drug approval reviews has occurred due to FDA modifications to the review process and the assessment of "user fees"—charges levied on sponsors for certain new drug and biologic applications—allowed under the Prescription Drug User Fee Act of 1992. FDA has used these funds to hire more reviewers to assess applications, and at the same time, the agency has strengthened the drug and biologic review processes to reduce review times. Similarly, the impetus for approval of new drugs to treat AIDS and other life threatening diseases has caused FDA to work toward improvement in the drug review process.

The documentation in the NDA addresses what happened during the clinical trials, how the drug is constituted (i.e., its components and composition), results of the prior animal studies, how the drug behaves in the body, and how it is manufactured, processed, and packaged. FDA also requires samples of the drug and its labels. If FDA's evaluation of the NDA reveals major deficiencies, substantially more work by the drug's sponsor may be needed, ranging from further analyses to conducting new clinical studies. In either case, the evaluation time for the NDA will be extended.

FDA classifies both INDs and NDAs to assign review priority on the basis of the drug's chemical type and potential benefit. The order in which applications are reviewed is determined with the aid of this classification system. The idea is to give priority to drugs with the greatest potential benefit.

In the final analysis, FDA's decision whether to approve a new drug for marketing centers on two basic questions: (1) do the results of well-controlled studies provide substantial evidence of effectiveness? and (2) do the results show the product is safe under the conditions for use in the proposed labeling? Safe, in this context, means that the benefits of the drug appear to outweigh its risks.

Recently, FDA has undertaken a number of ways to accelerate approval of drugs promising significant benefit over existing therapy for serious or life-threatening illness. These mechanisms incorporate elements aimed at making sure that rapid review and approval is balanced by safeguards to protect both public health and the integrity of the regulatory process. To provide greater speed and efficiency in the drug approval process, FDA has developed procedures to encourage collaboration between the drug sponsor and the FDA from the early preclinical phase through postmarketing surveillance. Central to this initiative is the establishment of ongoing clinical trial monitoring and evaluation by FDA, active consideration by the agency of treatment IND designation (discussed below) at the end of Phase II, and in some cases, the elimination of Phase III clinical trials. Similarly, FDA has allowed accelerated approvals to be used when approval can be reliably based on evidence of a drug's effect on a surrogate endpoint, i.e., a laboratory finding or physical sign that may not, in itself, be a direct measurement of how a patient feels, functions or survives but nevertheless is considered likely to predict therapeutic benefit. An example would be CD4 cell counts, used to measure the strength of the immune system. Under this type of accelerated approval program, FDA approves the drug on the condition that the sponsor continue to evaluate the actual clinical benefit of the drug.

FDA's treatment IND program provides a mechanism that allows promising investigational drugs to be used in expanded access protocols that allow relatively unrestricted studies in which the intent is both to learn more about the drug, especially its safety, and to provide treatment for people with immediately life-threatening or otherwise serious diseases for which there is no real alternative. Such expanded access protocols, however, still require researchers to investigate the drug in well-controlled studies and to supply sufficient evidence that the drug is likely to be helpful. The drug cannot expose patients to unreasonable risks.

FDA's parallel track initiative, outlined in a Public Health Service policy statement in April, 1992, establishes an administrative mechanism to expand the availability of promising investigational therapies beyond the parameters of the treatment IND regulations. Access to an investigational drug may be authorized as early as the end of Phase I when Phase II controlled clinical trials have been approved by FDA and patient enrollment for those trials has been initiated. The parallel track program is intended primarily for those patients who are unable to participate, for medical or geographic reasons, in clinical trials for the drug. FDA's parallel track policy is designed exclusively for patients with AIDS and HIV-related conditions.

Medical devices

Regulatory controls for medical devices depend primarily on a three-tiered device classification system established by the Medical Device Amendments of 1976 which substantially revised FDA's controls over devices under the FDCA. This statutory classification approach recognizes that medical devices vary widely in their simplicity or complexity and their corresponding degree of risk or benefit. In short, the statutory scheme underscores the fact that all devices do not need the same degree of regulation. Devices in Class I are those for which basic, general controls (such as adulteration, misbranding, registration, and recordkeeping requirements) alone are sufficient to assure safety and effectiveness. Devices in Class II are those for which general controls alone are insufficient to provide reasonable assurance of safety and effectiveness and for which sufficient information exists to establish special controls to provide this assurance. Special controls may include requirements such as mandatory performance standards, patient registries, or postmarket surveillance. Finally, devices in Class III are those for which insufficient information exists to assure that general controls and special controls provide reasonable assurance of safety and effectiveness. As a general rule, Class III devices are those represented to be life sustaining or life supporting, those implanted in the body, or those presenting potential unreasonable risk of illness or injury. Class III devices must have approved premarket approval applications (PMAs) before commercial distribution.

The premarket approval process for a Class III device is similar to that for new drugs. An investigational device exemption (IDE) allows device sponsors to conduct clinical studies on human subjects. FDA's IDE regulations (21 C.F.R. Part 812) distinguish between significant and nonsignificant risk device investigations, and the procedures for obtaining an IDE differ accordingly. The determination of whether a device investigation presents a significant risk is initially made by the device sponsor. A proposed study is then submitted to an Institutional Review Board (IRB) for review and approval of the clinical protocol. If the IRB agrees with the sponsor that the device study presents a nonsignificant risk, no IDE application to FDA is necessary before initiating the study. A sponsor of a significant risk device investigation must, however, obtain both IRB and FDA approval before beginning the study.

Following completion of these clinical studies, a premarket approval application (PMA) is submitted to FDA. The PMA must contain valid scientific evidence showing that there is reasonable assurance of the safety and effectiveness of the device under its conditions of use. Reasonable assurance of safety for devices is comparable to that for drugs, but reasonable assurance of efficacy may be based on well-controlled clinical studies, similar to the criteria for drugs, or from "valid scientific evidence" from which experts can fairly and reasonably conclude that the device will be effective for its intended uses. This statutory difference is intended for "meaningful" data developed under procedures less rigorous than well-controlled clinical investigations rather than anecdotal medical experience. FDA is now permitted to use information that demonstrates safety and effectiveness from previously submitted PMAs in evaluating a new PMA under certain conditions.

An incomplete PMA is not considered to be filed until it is resubmitted and is accepted for filing. If the PMA is suitable for scientific review, the filing date is the date FDA received the complete PMA, and the 180 day statutory review period provided by the FDCA begins on this filing date. Substantial changes may "restart" the "clock" when an amended PMA is filed. Once accepted for filing, the PMA undergoes scientific review by appropriate FDA personnel and advisory committees.

Class I or Class II devices are cleared for commercial distribution or marketing in the United States by a premarket notification process commonly referred to as a 510(k). In this process, the device manufacturer is required at least 90 days before commercial distribution of the device to give notice and provide certain information to FDA pursuant to § 510(k) of the FDCA. The submitter of this premarket notification must receive a letter or order from the agency permitting commercial distribution. Such an order is based on FDA finding the device substantially equivalent to one or more predicate devices already legally marketed in the United States for which premarket approval is not required. The manufacturer is obligated to provide in the 510(k) submission, among other things, evidence of such substantial equivalence.

What constitutes substantial equivalence is explained in detail in § 513(i)(1)(A) of the FDCA (21 U.S.C. 360c(i)(1)(A). Substantial equivalence means that a device has the same intended use and the same technological characteristics as the predicate device or has the same intended use and different technological characteristics, but it can be demonstrated that the device is as safe and effective as the predicate device and does not raise different questions regarding safety and effectiveness from the predicate device. Certain Class I devices are exempt by regulation from this 510(k) requirement. In certain situations, performance data from device testing will be necessary to demonstrate substantial equivalence. The need for detailed performance data, such as bench testing, animal studies, and/or clinical data, depends on the complexity of the device and the specific information needed to demonstrate equivalency. FDA does not normally require that the data be of the magnitude needed in a PMA to determine the safety and effectiveness of the device.

The 510(k) notification requirement applies: (1) whenever a manufacturer markets a device for the first time; (2) when there is a major change in the intended use of the device; or (3) whenever an existing device being marketed is modified in a way that could significantly affect its safety and effectiveness. In submitting a 510(k) to FDA, the manufacturer must include either a summary of the safety and effectiveness information upon which the substantial equivalence determination is based or state in the 510(k) that the safety and effectiveness information will be made available to anyone upon request.

Biological products

Premarket approval of biological products is required under Section 351(a) of the PHS Act (42 U.S.C. § 262). Licenses for biologics are issued upon a showing that the establishments and products "meet standards, designed to ensure the continued safety, purity, and potency of such products . . . ." (42 U.S.C. § 262(d)). A biological product's effectiveness for its intended use must be shown as part of the statutory requirement for potency (21 C.F.R. § 600.3(s)). At the investigational stages, when the products are being studied in clinical trials to gather safety and effectiveness data, biological products must meet IND requirements established in 21 C.F.R. Part 312. FDA regulation of biological products requires the submission of both product license applications (PLAs) and establishment license applications (ELAs) (21 C.F.R. § 601.1. *et seq.*). However, FDA is moving to a general biologic license application (BLA) for certain biological products that would combine elements of establishment and product licences. Biologics establishments and products must satisfy detailed standards set forth in FDA regulations (21 C.F.R. Parts 600–680). Moreover, there are statutory prohibitions against false labeling of biological products.

Although products regulated by FDA as biological products under the PHS Act are simultaneously also drugs or devices subject to applicable requirements in the FDCA, the agency does not require duplicate premarket approvals. For example, if FDA requires a PLA to be submitted for the product as a biologic, the agency does not also require submission of a new drug application (NDA) or a device premarket approval application (PMA).

COMBINATION PRODUCTS

Some products consist of a combination of biological products and drugs or devices. Under a provision of the Safe Medical Devices Act of 1990, which amended the FDCA, FDA determines the primary mode of action of the combination product and then assigns the primary jurisdiction for review of the product within the agency based on that determination. (21 U.S.C. § 353(g)). Under this statutory provision, FDA has established procedures for designating the organization within FDA (i.e., the Center for Biologics Evaluation and Research (CBER), the Center for Drug Evaluation and Research (CDER), or the Center for Devices and Radiological Health (CDRH)) to review combination products or any other products when the agency's center with primary jurisdiction is unclear. CBER, CDER, and CDRH have also entered into intercenter agreements to clarify the centers' responsibility for reviewing various kinds of products.

ENFORCEMENT

Both the FDCA and the PHS Act provide authority for enforcement of the statutory requirements. FDA is authorized to conduct inspections to determine compliance with regulatory requirements (21 U.S.C. § 360(h) and 374 and 42 U.S.C. § 262(c)). Premarket approvals for these products may be suspended or revoked (21 U.S.C. § 355(e) and 360e(e) and 42 U.S.C. § 262(a)). Biological products and devices may be recalled under certain circumstances (42 U.S.C. § 262(d)(2) and 21 U.S.C. § 360h). Court actions, including seizures of violative products, injunctions, and criminal prosecutions, may also be initiated by FDA (21 U.S.C. §§ 332, 333, and 334 and 42 U.S.C. § 262(f)).

CURRENT FDA GUIDANCE ON TISSUE ENGINEERING

In making regulatory determinations concerning products developed using tissue engineering, one must appreciate that there is no "bright-line" jurisdictional standard. Indeed, FDA's approach in this field has been complicated by the evolving nature of this new technology and its related therapies. With some notable exceptions, such as human heart valves, dura matter allografts, processed skin and bone products, umbilical cord vein grafts, and corneal lenticules which are classified by FDA as medical devices, other human tissues were not regulated by the FDA until recently. Moreover, if human tissue is processed for use only in a specific patient, especially when the cell or tissue processing is done by a physician or in a surgical setting, these activities are generally viewed as falling within the "practice of medicine" and are not regulated by FDA. If FDA considers a procedure or product used on a patient to constitute the "practice of medicine" no IND, IDE, or similar agency approval is required. Whether a particular procedure or practice falls within the "practice of medicine" exclusion is generally evaluated on a case by case basis. Decisions in this area are often difficult to distinguish; however, as a general rule, the extension of a medical procedure or the tissue-derived product to more than a single patient may limit the ability to claim that the activity is beyond regulation by FDA. Consequently, when a tissue-engineered technology is applied to a general patient population, it will likely engender FDA's involvement and require clinical investigation under FDA control.

FDA's efforts to regulate the products of tissue engineering are of recent vintage and rely on long-standing statutory authorities. Although FDA has worked to clarify its approach to regulation in this field, advancement of new cell and gene therapies continually strain the agency's resources and ability to provide useful guidance to those involved in research and development in this area. Beginning in 1983, FDA issued its first "Points to Consider" (PTC) document to address a variety of biotechnology topics, including monoclonal antibodies, recombinant DNA biological products, and interferon (FDA, 1984). More recent PTC guidances (FDA, 1996a) set forth FDA's position on a variety of topics central to the tissue engineering field. Although these PTC guidances are useful, it must be recognized that FDA has emphasized the need to reconsider periodically its approach in this evolving field. Indeed, FDA has stated that as novel therapeutic applications are developed and data accumulates about risk and benefits of this technology, the agency's regulatory approach may shift or need modification. Notwithstanding the evolving nature of the agency's regulatory approach, the guidance issued to date is instructive.

Human somatic cell therapy products and gene therapy

In a seminal statement on the regulatory policies appropriate for these new technologies, the FDA addressed somatic cell and gene therapy products in a 1991 PTC and an October, 1993 notice in the *Federal Register* (FDA, 1993). The agency defines somatic cell therapy products as autologous (self), allogeneic (intra-species), or xenogeneic (inter-species) cells that have been propagated, expanded, selected, pharmacologically treated, or otherwise altered in biological characteristic *ex vivo* to be administered to humans for the prevention, treatment, cure, diagnosis or mitigation of disease or injuries.

FDA's regulatory controls over somatic cell therapy products are based on a determination by the agency that cells manipulated in a way that changes the biological characteristics of the cell population will be subject to product licensure as final biological products. Examples include: (1) autologous or allogeneic lymphocytes activated and expanded *ex vivo;* (2) encapsulated autologous, allogeneic, or xenogeneic cells or cultured cell lines intended to secrete a bioactive factor or factors (*e.g.,* insulin or growth hormone); (3) autologous or allogeneic somatic cells (*e.g.,* hepatocytes, myocytes, fibroblasts, or lymphocytes) that have been genetically modified; (4) cultured cell lines; and (5) autologous or allogeneic bone marrow transplants using expanded or activated bone marrow cells (FDA, 1996c).

Those cells and tissues subject to biological product licensure as source material include allogeneic or xenogeneic cells harvested by other than the final product license holder and intended for manufacture into a somatic cell product. Examples in this category include: (1) muscle cells removed from donors and shipped to a manufacture for expansion into a muscle cell therapy; (2) animal cells harvested at an animal care facility and shipped to a manufacturer for encapsulation or other manufacturing steps into a somatic cell therapy; and (3) other human tissue harvested from donors and shipped to another entity for manufacturing into a somatic cell therapy.

The FDA guidance also focuses on gene therapy products which are defined as products containing genetic material administered to modify or manipulate the expression of genetic material or to alter the biological properties of living cells. Many gene therapy products, such as those containing viral vectors, will be regulated by FDA as biological products and will, thus, be subject to the licensing provisions of the PHS Act, as well as to the drug provisions of the FDCA. Other gene therapy products, such as chemically synthetized products, are considered drugs and are regulated under the relevant drug provisions of the FDCA only.

Regulatory controls governing somatic cell and gene therapy products focus on certain basic criteria. First, FDA must be satisfied that these products are derived from sources that are well characterized, uniform, and not contaminated by hazardous adventitious agents. Second, rigorous control of the manufacturing process is essential because of difficulties inherent in assessing the consistency of biologic products. Finally, the testing of biologic potency is given particular emphasis and may require the development of new methodologies.

Somatic cell and gene therapy products are also subject to other pertinent regulatory requirements, including provisions governing drug listing and regulation, and rules governing misbranding and adulteration. FDA has acknowledged that applications for premarket approval are not required at present for certain other cellular products, including minimally manipulated or purged bone marrow and certain minimally processed cell transplants.

The manufacture of somatic cell therapy products or gene therapy products may involve ancillary products, such as culture cell medium and substances used to activate or modify the biological characteristics of cells, used as part of the manufacturing process. A common characteristic of ancillary products is that they are intended to act on cells, rather than to have an independent effect on the patient. The ancillary products are not intended to be present in the final product but may have an impact on the safety, purity, or potency of the product under manufacture. Such ancillary products meet the definition of devices and, when marketed, will be regulated under the device provisions of the FDCA, with the level of regulatory control being based on the medical device classification scheme outlined earlier. In contrast, products administered directly to patients or products whose function requires incorporation into the somatic cells are not considered ancillary products; rather, they will be regulated as drugs or biologics. Moreover, when these ancillary products are used to manufacture somatic cell or gene therapy products, they become subject to drug CGMPs, in particular for components and containers. Manufacturers who wish to market ancillary products for use in the manufacturing of somatic cell or gene therapy products must file either: (1) a 510(k), (2) a PMA, or (3) an amendment to an existing 510(k), PMA, NDA, or PLA.

Products containing both a somatic cell component and another drug or device component in the final product will be regulated as combination products. Some of the complex regulatory issues represented by cellular and gene therapies is illustrated by pancreatic islet cells secreting insulin. If the cells are in an inert polymeric matrix designed to prevent their rejection upon implantation, the matrix is a device, the secreted insulin is a nonbiological drug, and the islet cells are a biologic. FDA in a 1996 draft addendum to its 1991 PTC and 1993 guidance, opined that the regulatory controls governing clinical trials for these emerging therapies should not unnecessarily impede their development. Indeed, FDA has suggested that in exploratory Phase I trials for somatic cell and gene therapy products, not as much data are needed prior to beginning Phase I as would be needed for advanced product development, especially in the case of therapies for severe or life-threatening diseases. On the other hand, data sufficient to support Phase I trials, may not be adequate at later stages.

In this regard, IND applications for somatic cell and gene therapies should follow the basic format and content for INDs for any investigational biological product. The core of the IND should provide information on the general investigational plan, the clinical protocol, a description of previous human experience, and a patient consent form. Moreover, IND applications for such products should contain background information on the proposed therapy, including a detailed description of the somatic cell or gene therapy product itself and the expected effects of its expression and a rationale for the therapeutic intervention. Other areas that should be covered in the

IND include a description of the source materials for the products, including characterization and quality control of the master and working cell banks and virus stocks, manufacturing and purification steps, stability and potency of the product, and preclinical pharmacological and toxological studies.

FDA regulations affecting the recovery, processing, storage, or distribution of banked human tissue

FDA issued an interim final rule in December, 1993 (21 C.F.R. Part 1270), imposing certain regulatory requirements on the recovery, processing, storage, and distribution of human tissue intended for administration to another human for the diagnosis, cure, mitigation, treatment, or prevention of any condition or disease. These regulations impose controls over infectious disease screening and testing on humans who are the source of tissue for transplantation. The regulations, however, do not apply to human tissue manipulated or processed by methods that change tissue function or characteristics. Moreover, these requirements for banked human tissue do not apply to products currently regulated by FDA as drugs, biological products, or medical devices. Vascularized human organs and semen or other reproductive tissues, human milk, bone marrow, and autologous products are also excluded. Thus, although by definition, the interim final rule does not have application to most of the products derived from tissue engineering, it demonstrates FDA's interest in expanding its regulatory control over all areas associated with the use of tissue for transplantation in humans.

Placental/umbilical cord blood stem cell products

In December, 1995, FDA released a draft guidance document for the regulation of placental/umbilical cord blood stem cell products (hereinafter referred to as cord blood stem cell products) intended for transplantation or further manufacture into injectable products (FDA, 1995). FDA defines cord stem cell products as autologous or allogeneic products that contain hematopoietic stem cells derived from placental/umbilical cord blood and are administered to humans for the prevention, treatment, cure, diagnosis, or mitigation of disease or injuries. FDA concluded that the safety, potency, and purity of these products can only be assured at this time by regulating these products as biologics, pursuant to the PHS Act. At the same time, the agency acknowledged that these products also fall within the definition of drugs and thus are subject to regulations applicable to drugs promulgated under the FDCA. As biological products, cord blood stem cell products are subject to requirements for PLAs and ELAs.

Cord blood stem cells intended for use as source material for further manufacture into licensed hematopoietic stem cell products will also be licensed as biological products, unless they are already regulated under an ELA for the final product. This treatment by the agency recognizes that cord blood stem cells intended for further manufacture may be part of a cooperative manufacturing agreement in which two or more manufacturers perform different aspects of the manufacture of a product that ordinarily warrants separate licensing. In this situation, FDA will accept only license applications for biological products intended for further manufacture that specify the licensed manufacturer to which the intermediate product will be shipped. FDA will approve such applications only after safety and efficacy of the end product has been demonstrated.

FDA's recent guidance also addresses ancillary products used as part of the manufacturing process for cord blood stem cell products. Although these products are not intended to be present in the final product, they may have an impact on the safety, purity, or potency of the final product. Consequently, ancillary products that are not already regulated under an existing IND, NDA, PLA, PMA, or premarket notification 510(k) will be classified as devices. When such ancillary products are used in the manufacture of cord blood stem cell products, their use will be subject to either drug or device CGMPs. Manufacturers who wish to market ancillary products separately for use in the manufacturing of hematopoietic stem cell products must file either a 510(k), a PMA, or an amendment to an existing 510(k), PMA, NDA, or PLA depending on the type of ancillary products.

Peripheral blood hematopoietic stem cell products

In February 1996, FDA issued draft guidance on the regulation of peripheral blood hematopoietic stem cell products (hereafter referred to as peripheral blood stem cell products) intended for transplantation or as source material for further manufacture into injectable products (FDA, 1996b). Peripheral blood stem cell products are defined as hematopoietic cells derived from peripheral blood to be administered to humans for the prevention, treatment, cure, diagnosis, or mitigation of disease or injuries.

These products can be divided into two groups based on the degree of manipulation necessary to produce the final product. Manipulated peripheral blood stem cells are products obtained after one or more procedures have been performed to purge or enrich the starting material of a subset(s) of nucleated cells. Such products also include products subject to licensure as a somatic cell therapy as previously described in FDA's 1993 guidance on human somatic cell therapy. Examples of procedures that result in manipulated products include centrifugal elutriation, negative or positive cell selection by monoclonal antibody-based technologies, cell populations expanded *in vitro* using cytokines or other procedures leading to a somatic cell therapy product. In contrast, nonmanipulated or minimally manipulated peripheral blood stem cells are products that have not been subjected to a procedure that selectively removes, enriches, expands, or functionally alters specific nucleated cell populations.

Consistent with the regulatory approach for cord blood stem cell products, FDA has stated that the safety, potency, and purity of manipulated peripheral blood stem cells is best assured at this time by regulating these products as biologics. As such, the safety and efficacy of these products will be evaluated under an IND and the resulting final product licensed as a biologic. FDA does not expect the submission of IND applications for hematopoietic stem cells that are isolated from peripheral blood without manipulation. However, if these products are intended for distribution in interstate commerce, licensure of the product and establishment is required. Indeed, FDA requires manufacturers of these products to register as blood establishments (21 C.F.R. 607). They must operate according to applicable blood CGMPs and are subject to FDA inspection. Nonmanipulated peripheral blood stem cells can be licensed when they meet the product specifications and standards for collection, processing, and storage. These standards are at present under development by FDA.

Nonmanipulated peripheral blood stem cells intended as a source material for the manufacture of a final hematopoietic stem cell product are also subject to licensure unless regulated under the license for the final product. Products to be utilized as source materials will only be approved when the final product (such as the genetically altered cell) is approved. Examples include peripheral blood stem cells intended for *ex vivo* expansion or as targets of gene transfer and expression. Evaluation of source material that is manufactured differently from accepted standards, may be subject to IND regulation as well.

Ancillary products used in the production of peripheral blood stem cell products will be regulated as medical devices. A common characteristic of these products is that they are intended to act on the cells rather than to have an independent effect on the patient. The intended action of such products is not dependent upon incorporation into the peripheral blood stem cell product.

In contrast, products administered directly to patients or products whose function requires incorporation into the hematopoietic stem cell product to maintain structural or functional integrity are not considered medical devices; rather, they will be regulated as drugs or biological products. Examples include investigational agents administered to stem cell donors to mobilize the stem cell populations before collection of the product and storage medium and cryoprotective agents added to the stored product and infused with the product into the patient.

Products comprised of living autologous cells manipulated *ex vivo* and intended for implantation for structural repair or reconstruction

Perhaps in recognition of the evolving field of cell and gene therapy products, FDA announced in March 1996 its intention to hold public meetings on the good manufacturing practices and clinical issues related to products comprised of living autologous cells manipulated *ex vivo* and intended for surgical repair or reconstruction. By seeking public comment on the most appropriate regulatory scheme for such products, FDA seems to be expressing a willingness to reassess critical areas relating to the development and production of these products. This may signal FDA's readiness to consider innovative regulatory approaches for the array of emerging therapies in the tissue-engineering field. This dialogue could lead to the implementation of less strenuous or more streamlined pathways to regulatory approval.

For example, it appears likely that FDA will initially impose case-by-case reviews of manipulated autologous cell products. Although a final regulatory strategy has not emerged, FDA has recently signaled a readiness to avoid long-term clinical studies by allowing the use of short term (one year or less) endpoints directly measuring clinical benefits to demonstrate efficacy in cases where a favorable risk-benefit evaluation has been established and long-term safety concerns are low. Under this approach, FDA would accept postmarketing studies to assess long-term safety and efficacy and allow a more flexible approach to current good manufacturing practices.

References

Code of Federal Regulations Title 21, Parts 201, 202, 207, 211, 312, 314, 600–680, 807, 812, and 820
Federal Food, Drug and Cosmetic Act, 21 U.S.C. §§ 301 et seq
Food and Drug Administration (1993): Application of Current Statutory Authorities to Human Somatic Cell Therapy Products and Gene Therapy Products. *Fed Reg* 58:53248
Food and Drug Administration (1984): Biological Products; in Vitro or in Viro Monoclonal Antibodies, Products Made Using Recombinant DNA Technology, or Interferon. *Fed Reg* 49:1138
Food and Drug Administration, Center for Biologics Evaluation and Research, Draft Document Concerning The Regulation Of Placental/Umbilical Cord Blood Stem Cell Product Intended For Transplantation Or Further Manufacture Into Injectable Products (1995)
Food and Drug Administration, Center for Biologics Evaluation and Research, Draft Addendum To The Points To Consider In Human Somatic Cell And Gene Therapy (1996a)
Food and Drug Administration, Center for Biologics Evaluation and Research,

Draft Document Concerning The Regulation Of Peripheral Blood Hematopoietic Stem Cell Products Intended For Transplantation Or Further Manufacture Into Injectable Products (1996b)

Food and Drug Administration (1996c): Regulatory Approach To Products Comprised of Living Autologous Cells Manipulated Ex Vivo And Intended For Structural Repair Or Reconstruction. *Fed Reg* 61:9185

Jacobson E, Johnson G (1991): Regulating tissues (and organs?) as devices. *Food Drug Cos LJ* 46(spec iss) 25

Kessler D, Siegel J, Noguchi P, Zoon K, Feiden K, Woodcock J (1993): Regulation of somatic-cell therapy and gene therapy by the Food and Drug Administration. *NEJM* 329:1169

Korwek E (1995): Human biological drug regulation: Past, present, and beyond the year 2000. *Food Drug LJ* 123(spec iss)

Parkman P, Noguchi P (1991): Regulating tissues (and organs?) as biologics. *Food Drug Cos LJ* 46(spec iss):21

Porter M, Buc N (1991): Legal base of tissue and organ regulation. *Food Drug Cos LJ* 46(spec iss): 11

Public Health Service Act, 42 U.S.C §§ 201 et seq

4

SYNTHESIS AND PROPERTIES OF BIODEGRADABLE POLYMERS USED AS SYNTHETIC MATRICES FOR TISSUE ENGINEERING

WAI HUNG WONG AND DAVID J. MOONEY

INTRODUCTION

Motivation for tissue engineering

Organ or tissue failure remains a frequent, costly, and serious problem in health care despite advances in medical technology. The large number of patients suffering from tissue losses or organ failures is demonstrated by the approximately 8 million surgical procedures and 40 to 90 million hospital days required annually to treat these problems (Langer and Vacanti, 1993). Available treatments now include transplantation of organs from one individual to another, performing surgical reconstruction, use of mechanical devices (e.g., kidney dialyzer), and drug therapy. However, these treatments are not perfect solutions. Transplantation of organs is limited by the lack of organ donors, possible rejection, and other complications. For example, there are only 3000 donors available in contrast to 30,000 patients who die from liver failure each year (American Liver Foundation, 1988). Mechanical devices cannot perform all functions of an organ, e.g., kidney dialysis can only help remove some metabolic wastes from the body. Likewise, drugs can only replace limited biochemical functions of an organ or tissue, and physiologic control of drug levels comparable to the control systems of the body is difficult to achieve. For example, although diabetes can be partially treated by administration of insulin, it is still the leading cause of chronic renal failure and blindness (Chukwuma,

Synthetic Biodegradable Polymer Scaffolds
Anthony Atala and David Mooney, Editors
Robert Langer & Joseph P. Vacanti, Associate Editors
© 1997 Birkhäuser Boston

1995; Hoelscher et al, 1995; Sander and Wilson, 1993). This is partly due to difficulties in controlling the drug level *in vivo* (Levesque et al, 1992). Finan-
cially, the cost of surgical procedures is very high ($150,000 for a liver
transplant). It is estimated that almost half of the United States $800 billion in
annual health care costs are due to loss or malfunction of tissue or organs
(Langer and Vacanti, 1993).

Advances in medical, biological, and physical sciences have enabled the
emergence of the field of tissue engineering. "Tissue engineering" is the
application of the principles and methods of engineering and the life sciences toward the fundamental understanding of structure/function relation-
ships in normal and pathological mammalian tissues and the development
of biological substitutes to restore, maintain, or improve function (Nerem
and Sambanis, 1995). It thus involves the development of methods to build
biological substitutes as supplements or alternatives to whole organ or tissue
transplantation. The use of living cells and/or extracellular matrix (ECM)
components in the development of implantable parts or devices is an attractive approach to restore or to replace function. The advantage of this approach over whole organ/tissue transplantation is that only the cells of
interest are implanted, and they potentially can be multiplied *in vitro*. Thus,
a small biopsy can be grown into a large tissue mass and, potentially, can be
used to treat many patients. The increased tissue supply may reduce the cost
of the therapy because early intervention is possible during the disease, and
this may prevent the long-term hospitalization that results as tissue failure
progresses. The use of immunosuppression may also be avoided in some
applications by using the patient's own cells.

Importance of matrices in tissue engineering

Matrices play a central role in tissue engineering. Matrices are utilized to
deliver cells to desired sites in the body, to define a potential space for the
engineered tissue, and to guide the process of tissue development. Direct
injection of cell suspension without matrices has been utilized in some cases
(Brittberg et al, 1994; Matas et al, 1976; Ponder et al, 1991), but it is difficult
to control the placement of transplanted cells. In addition, the majority of
mammalian cell types are anchorage dependent and will die if not provided
an adhesion substrate. Polymer matrices can be used to achieve cell delivery
with high loading and efficiency to specific sites (Mooney et al, 1995b;
Vacanti et al, 1988). Polymer matrices also provide mechanical support
against compressive and tensile forces, thus maintaining the shape and integrity of the scaffold in the aggressive environments of the body. The
scaffold may act as a physical barrier to immune system components of the
host or act as a matrix to conduct tissue regeneration, depending on the
design of the scaffold. The first type of scaffold, immunoprotective devices,

utilize a semipermeable membrane to limit communication between cells in the device and the host. The small pores in these device (d < 10 Nnm) allow low molecular weight proteins and molecules to be transported between the implant and the host tissue, but they prevent large proteins (e.g., immuno-globulins) and host cells (e.g., lymphocytes) of the immune system from entering the device and mediating rejection of the transplanted cells. In contrast, open structures with large pore sizes (d > 10 μm) are typically utilized if the new tissue is expected to integrate with the host tissue (Mooney and Langer, 1995). The morphology of the matrix can guide the structure of an engineered tissue (Vacanti et al, 1988), including the size, shape, and vascularization of the tissue (Mooney et al, 1995b).

The ideal matrix should also elicit specific cellular functions and direct cell–cell interactions. Tissue engineering matrices function as synthetic ex-tracellular matrices, and the proper design of these matrices may allow them to mimic the entire range of mechanical and biological functions of the native ECM. Biomimetic synthetic polymers armed with cell-adhesion pep-tides can be synthesized as implants for tissue regeneration or cell transplan-tation to achieve this purpose (Hubbell, 1995).

A variety of naturally occurring and synthetic polymers can be used to fabricate tissue engineering matrices. Naturally derived materials (e.g., hyaluronic acid and collagen in ECM) must be isolated from plant, animal, or human tissues and are typically expensive and suffer from large batch-to-batch variations. Synthetic polymeric materials, on the other hand, usually can be precisely controlled in material properties and quality. Moreover, syn-thetic polymers can be processed with various techniques and supplied con-sistently in large quantities. The mechanical and physical properties of synthetic polymers can be readily adjusted through variation of molecular structures in order to fulfill their functions without the use of either fillers or additives. Table 1 outlines different structural factors of polymers that can be used to adjust a variety of critical properties. The use of biodegradable syn-thetic polymers in tissue engineering is especially attractive. The bioresorp-tion or disintegration of these materials after they have fulfilled their function minimizes the chronic foreign body response and leads to the formation of a completely natural tissue. This chapter will focus on the synthesis and proper-ties of biodegradable synthetic polymers for these reasons.

Types of tissue engineering matrix materials

A variety of synthetic biodegradable polymers can be utilized to fabricate tissue engineering matrices. In general, these materials may be utilized as structural elements in the scaffold, to deliver or immobilize the cells, or to achieve both purposes. Poly(glycolic acid) (PGA) and poly(lactic acid) (PLA) (Figure 1) are by far the most commonly used synthetic polymers in

Table 1. Structural variables used to control biodegradable polymer properties.

Variables	Effects	Examples
Incorporation of both natural and/or non-natural monomers	May reduce/eliminate immunologic response often found in naturally-derived polymers	Nonimmunologic PGA and PLA (vs. collagens)
Incorporation of labile groups in polymer chain	Control kinetics of biodegradation	Hydrolyzable ester bond in PGA
Incorporation of functional groups in side chains	Control chemical and physical properties of polymers	Hydrophilic, hydrophobic and amphiphilic polyphosphazenes
Incorporation of chiral centers in polymer chains	Control physical and mechanical properties of polymer	Semi-crystalline I-PLA and amorphous dl-PLA
Possibility of utilizing multiple monomers	Control properties of polymers	Glycolic and lactic acids in PLGA
Use of natural compounds as monomers	Biocompatible breakdown products	Lactic acid in PLA
Use of different polymer architectures	Control physical and mechanical properties of polymersexhibits	Branched polymers lower viscosity than linear ones

tissue engineering. These polymers are also extensively utilized in other biomedical applications such as drug delivery. They have a long history of use in humans since their introduction as suture materials in the 1970s and are FDA approved for a variety of applications (Huang, 1989). Therefore, the advantages and disadvantages of this class of polymers as structural materials for tissue engineering will be discussed in more detail in this chapter. However, in principle, any biodegradable polymer that produces nontoxic degradation products can be used in tissue engineering. The polymers listed in Table 2 are usually recognized as biodegradable. We will discuss the chemistry and structure of a selected number of these polymers to give the reader an idea of specific tissue engineering applications for which these polymers might be suitable.

Hydrogels, either comprised of synthetic polymer or natural polysaccharides, have been utilized to immobilize transplanted cells. Natural polysaccharides and proteins can be either chemically modified or combined with synthetic elements to induce desirable properties for specific applications. These materials will also be discussed to develop an appreciation for their potential as tissue engineering matrices.

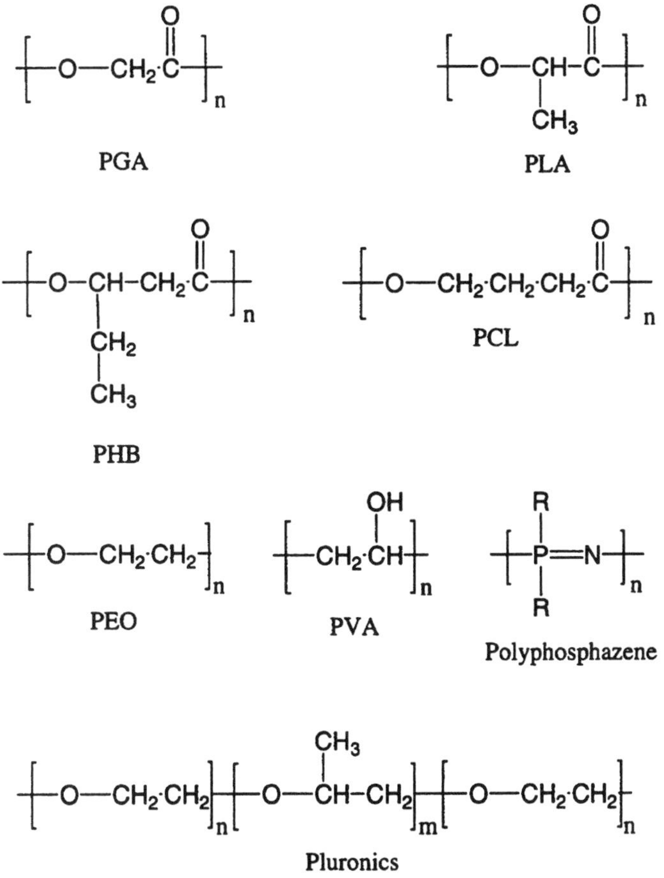

Figure 1. Chemical structures of poly(glycolic acid) (PGA), poly(lactic acid) (PLA), poly(β-hydroxbutyrate) (PHB), poly(caprolactone) (PCL), poly(ethylene oxide) (PEO), pluronics, and poly(phosphazene).

BIODEGRADABLE POLYMER MATRICES

Structural materials

PGA and PLA

Aliphatic polyesters of the poly(α-hydroxy acids) are perhaps the most common biodegradable synthetic polymers known. Their general formula is -[–O–CH(R)–CO–]- which derives from corresponding HO–CH(R)–COOH where R = H in the case of glycolic acid (GA) and R = CH₃ in the

Table 2. Main polymers recognized as biodegradable

Synthetic
  Polypeptides
  Polydepsipeptides
  Nylon-2/nylon-6 copolyamides
  Aliphatic polyesters
  - Poly(glycolic acid) (PGA) and copolymers
  - Poly(lactic acid) (PLA) and copolymers
  - Poly(alkylene succinates)
  - Poly(hydroxy butyrate) (PHB)
  - Poly(butylene diglycolate)
  - Poly(ε-caprolactone) and copolymers
  Polydihydropyrans
  Polyphosphazenes
  Poly(ortho ester)
  Poly(cyano acrylates)

Natural
  Modified polysaccharides (cellulose, starch, chitin, etc.)
  Modified proteins (collagen, fibrin, etc.)

Adapted from Vert and Guerin, 1991.

case of lactic acid (LA), the latter being chiral, i.e., D- or L-isomer is possible. These polymers have been used extensively in bone osteosynthesis and reconstruction, as reviewed by Vert et al (1984), and later in drug delivery as reviewed by Gombotz and Pettit (1995). They have also been utilized in treating renal trauma (Yerdel et al, 1993) and pulmonary surgery (Naka-mura et al, 1990). There are also extensive examples using PGA, PLA, or their copolymers in tissue engineering of skin (Cooper et al, 1991; Hansbrough et al, 1992), liver (Kim and Vacanti, 1995; Johnson et al, 1994; Mooney et al 1995b, 1996), cartilage (Freed and Vunjak-NovaKovic, 1993; Ma et al, 1995; Puelacher et al, 1994), nerve (Merrell et al, 1986; Pham et al, 1991), and trabecular bone (Thomson et al, 1995).

Synthesis

PGA and PLA can be prepared by two different routes, namely, polycon-densation and ring opening polymerization (ROP). Generally, the simple polycondensation is less expensive, but the resulting polymers have low, uncontrolled molecular weight, and it is difficult to prepare copolymers (Gilding and Reed, 1979; Kricheldorf and Kreiser-Saunders, 1996). It is believed that the antimony trioxide catalyst typically used to effect polycon-densation acts as both polymerizing and depolymerizing agent. Moreover, glycolic and lactic acids have a great tendency to cyclodimerize under these conditions, and this renders simple polycondensation an unsuitable method. The preferred method for producing high molecular weight polymers is ROP of the cyclic dimer, glycolide (and/or lactide). Depending on the catalyst involved, three different mechanisms have been reported: cationic, anionic, and insertion. Among these, the insertion mechanism using metal alkoxides or carboxylates is the most desirable pathway and is the choice in commercial production (Frazza and Schmitt, 1971). Typical examples of catalysts of this class are aluminum, zinc, titanium, zirconium, antimony, tin(IV), and tin(II) alkoxides or carboxylates. The insertion mechanism allows the preparation of high molecular weight polylactides without racemization up to temperatures above 150°C (Kricheldorf and Kreiser-Saunders, 1996). The mechanism of ROP of lactones has been reviewed (by Penczek and Kubisa, 1989; Penczek and Slomkowski, 1989). The tin catalyst, tin(II) octoate, was used extensively because of its acceptance by the FDA as a food stabilizer. The polymerization of glycolide can be carried out in bulk at 220°C for 4 h, at which time a 96% conversion and molecular weights from 10^4 to 10^6 have been reported. Copolymerization of glycolide with lactide has also been investigated. The reactivity ratios at 200°C have been found to be 2.8 for glycolide and 0.2 for lactide. This indicates that copolymers of glycolic and lactic acids have broad compositional ranges, with glycolide always being preferentially polymerized at low conversions and lactide being incorporated to an ever-increasing extent as the glycolide is depleted (Gilding and Reed, 1979).

During the advanced stages of most ROPs, additional reactions such as ester–ester interchange and chain unzipping may take place. The extents of these reactions are affected by the reactivity of the ester moieties. These events can have a significant effect on the composition of the final product. Due to the ester exchange, cyclic dimer, trimer, and, to a less extent, cyclic oligomers can be found along with the reformed monomer. In the case of copolymerization, additional randomization of the polymer chain will occur as a consequence of the ester-ester exchange of different ester moieties (Shalaby and Johnson, 1994). The microstructures of PLGA copolymers can be determined by both proton and carbon NMR spectroscopy (Kasperczky, 1996). It has been reported that the block lengths increase and, at the same time, the extent of transesterification decreases with decreasing polymerization temperature.

Properties of PGA and PLA that affect their applications in tissue engineering

Mechanical strength and morphology

PGA was first developed as the synthetic absorbable suture, Dexon (Frazza and Schmitt, 1971). PGA has high crystallinility, a high melting point, and

Table 3. Crystallinity and thermal properties of PGA, PLA and copolymers

	% crystallinity	Tm	Tg
PGA	46–52	225	36
90:10 PGLA	40	210	37
50:50 PGLA	0	none	55
PLA	37	185	57
dl-PLA	0	none	N/A

N/A = not available. Adapted from Gilding and Reed, 1979.

low solubility in organic solvents (Table 3). The polymer (fiber grade, inherent viscosity = 1.2–1.6 dL/g in hexafluoro-isopropanol) can be spun into multifilament or monofilament yarns for the production of braided and monofilament sutures, respectively (Frazza and Schmitt, 1971; Chujo et al, 1967). A typical suture braid has a tensile strength of 80–100 Kpsi (Table 4). Owing to the hydrophilic nature of PGA, Dexon sutures tend to lose their mechanical strength rapidly (50%) over a period of 2 weeks and are absorbed in about 4 weeks after implantation (Frazza and Schmitt, 1971; Katz and Turner, 1970; Reed and Gilding, 1981).

The presence of an extra methyl group poly(L-lactic acid) (PLA) or poly(D-lactic acid) (d-PLA) makes them more hydrophobic than PGA. For instance, films of PLA only take up approximately 2% water (Gilding and Reed, 1979). In addition, the ester bond in PLA is less labile to hydrolysis due to steric hindrance of the methyl group. Therefore, PLA degrades much more slowly than PGA (Reed and Gilding, 1981) and has higher solubility in organic solvents. PLA is employed much more often than d-PLA, since the hydrolysis of PLA yields L-lactic acid which is the naturally occurring stereoisomer of lactic acid. Whereas PLA possesses about 37% crystallinity, the optically inactive poly(DL-lactic acid) (dl-PLA) is amorphous. The difference in the crystallinity of dl-PLA and PLA has important practical consequences. For instance, the amorphous dl-PLA is usually considered in drug delivery application in which a homogeneous dispersion of the active

Table 4. Mechanical properties of PGA and 90:10 PGLA

	PGA (Dexon)	90:10 PGLA (Vicryl)
Tensile strength (Kpsi)	106	95
Knot strength (Kpsi)	65	63
Elongation (%)	24	25

Adapted from Shalaby and Johnson, 1994.

species within a monophasic matrix is desired (Engelberg and Kohn, 1991). However, the semicrystalline PLA is preferred in cases where high mechanical strength and toughness are required, for example, in orthopedic devices (Hay et al, 1988; Leenslag et al, 1987; Vainionpaa et al, 1987). It is pertinent to note that γ-irradiation of PLA causes chain scission, cross-linking, and a decrease in crystallinity (Gupta and Deshmuth, 1983). Therefore, caution must be taken when sterilizing the polymer matrices by γ-irradiation.

To widen the range of materials properties exhibited by PGA, copolymers of GA and LA (PLGA) have been studied. Whereas PGA is highly crystalline, PLGAs usually exhibit lower crystallinity and Tm (Gilding and Reed, 1979). For example, while PGA and PLA are partially crystalline, 50:50 PLGA is entirely amorphous (table 3). These morphological changes result in an increase in the rates of hydration and hydrolysis. Thus, copolymers tend to degrade more rapidly than PGA and PLA. (Mooney et al, 1995a).

Biodegradation

The degradation mechanism of PGA and copolymers *in vitro* is usually regarded as bulk erosion (Gombotz and Pettit, 1995). This is evident from the fact that a significant molecular weight decrease usually precedes monomer release from the polymer samples (Figure 2). This mechanism of degra-

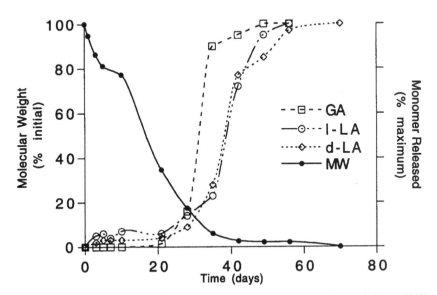

Figure 2. Change of molecular weight (MW) of devices fabricated from 50/50 PLGA, and release of glycolic (GA), d-lactic (d-LA), and l-lactic (l-LA) acids. Samples were placed in an aqueous medium at 37°C *in vitro*. (Adapted from Mooney et al, 1995a)

dation may be undesirable in certain applications. The relatively rapid release of large quantities of acid (glycolic and/or lactic acid) may lead to a local acidosis if a large mass of these polymers is present in a concentrated form (e.g., a solid pin). However, highly porous scaffolds are typically utilized in tissue engineering applications and contain a relatively low mass of polymer per unit volume (Figure 3). The highly porous structure of the scaffolds assists cell penetration as well as polymer degradation (Mooney et al, 1995a). The rate of degradation is affected by the morphology of the scaffold, and the large surface areas speed up the diffusion of water molecules into the bulk of the polymers when they are placed in an aqueous environment (e.g., *in vivo*). The polymers undergo random chain scission by simple hydrolysis of the ester bond linkage, and the monomer diffuses out of the polymer bulk into water (Reed and Gilding, 1981). It is important to note that loss of mechanical strength of PGA is faster when the polymer is incubated at a temperature higher than its Tg. This indicates that the glassy state protects PGA from hydrolysis since all short term chain motions are frozen. Water diffusion, and therefore hydrolysis, is more facile at tempera-

Figure 3. A SEM photomicrograph of a highly porous sponge fabricated from PLA (Mooney DJ, unpublished results). These types of matrices have been utilized to transplant a variety of cell types, including hepatocytes. (Mooney et al, 1995b)

tures above Tg. It is also relevant to mention that the Tgs of PGA and some copolymers are very close to the physiological temperature. Polymeric materials may undergo significant structural change after implantation due to water penetration and loss of mechanical strength. It is also speculated that enzymatic action may partially contribute to biodegradation of PGA *in vivo*.

The chemical compositions and the ratio of monomers used in the polymerization reaction strongly influence the degradation characteristics of the copolymer. The degradation rates for copolymers of GA and LA have been shown to be influenced by factors that affect polymer chain packing, i.e., crystallinity, and hydrophobicity. Since degradation is induced by hydrolysis, a crystalline structure or hydrophobic polymer composition disfavors dissolution and degradation. Gombotz and Pettit (1995) summarized the specific factors affecting copolymer crystallinity and hydrophobicity as follows: (1) the ratio of lactide to glycolide monomer in the copolymer; (2) the stereoregularity of the monomer units in the polymer affects polymer chain packing; (3) randomness of lactide and glycolide decrease the ability of chains to crystallize; and (4) low molecular weight polymers degrade faster than high molecular weight polymers, especially when the end groups are free acid rather than capped with ester or other groups. Some of these structural effects are demonstrated in Figure 4. Mass loss from polymer samples comprised of PLA is insignificant in the experimental time period ($\approx$50 weeks). However, those comprised of copolymers of GA and LA or dl-PLA degrade

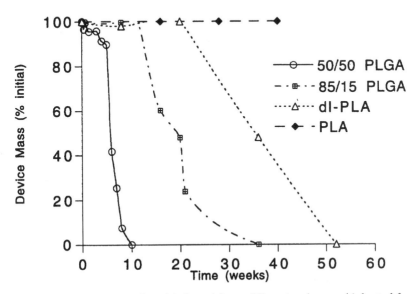

Figure 4. Mass loss of devices fabricated from different polymers (Adapted from Mooney et al, 1995a)

much faster (the higher the glycolic acid content, the higher the degradation rate) (Mooney et al, 1995a).

The presence of monomers and low molecular weight cyclic oligomers in absorbable polymers should be avoided, for they degrade much more rapidly than the polymers and can lead to undesirable chemical and biological effects (Shalaby and Johnson, 1994). It has been shown that polylactide with increased monomer content exhibits a higher rate of bioabsorption and a more drastic decrease of molecular weight (Nakamura et al, 1989).

Other chemistries for tissue engineering

Chemical modifications of PGA/PLA

Little modification of these polymers is possible because there are no other functional groups on the side chain, except the methyl of the lactic acid residue. One possibility to modify the properties of these polymers is to form copolymers with residues having more diverse side chain structures (e.g., lysine). A new monomer, 3-(Nε-benzoxycarbonyl-L-lysyl)-6-L-methyl-2,5-morpholinedione, was bulk copolymerized with L,L-lactide in the presence of stannous octoate as catalyst using the same ROP techniques utilized for lactide and glycolide (Barrera et al, 1993). The lysine content was determined by NMR technique to be approximately 1.3 mole %. A poly(lactide-co-lysine) functionalized with peptide containing the argininine-glycine-aspartate (RGD) sequence was prepared by removal of the benzyoxycarbonyl protecting group on the lysyl residue and peptide coupling. The peptide concentration was found to be approximately 3.1 mmol/g which could be translated into a peptide surface density of 310 fmol/cm^2. A surface density of as low as 1 fmol/cm^2 of an RGD peptide had been previously determined to promote cell adhesion to an otherwise nonadherent surface (Massia and Hubbell, 1991). Therefore, by carefully processing the copolymer, biodegradable films with cell adhering properties could be prepared from the copolymer of lactide and lysine.

Other strategies have also been employed to widen the properties of polylactides. For example, PLA has also been synthesized as an acrylic macromonomer and subsequently copolymerized with polar acrylic monomers (e.g., 2-hydroxyethylmethacrylate) (Barakat et al, 1996). These polymers were studied as amphiphilic graft copolymers for drug delivery purposes. The surface properties of these polymers may be controlled by the ratio of the PLA graft length and copolymer content and can be potentially used to control the drug release profile and biodistribution. Other examples of this approach include grafting PLA blocks to geraniol and pregnenolone (Kricheldorf and Kreiser-Saunders, 1996). However, even with such copolymerization techniques, the scope of chemistry that can be done is still limited.

Other polyesters

Properties of polyesters can also be varied by changing the structures of the polymer backbones. Polycaprolactone (PCL) (Figure 1), having two more carbon atoms than PGA on the polymer backbone, has been studied as a substrate for biodegradation and as a matrix for drug release systems (Huang, 1989). Its degradation *in vivo* is much slower than PGA; therefore, it is suitable for controlled release devices with long *in vivo* life times (1–2 years). PCL can be prepared by anionic ring opening polymerization of ε-caprolactone using metal hydroxide initiators (Jerome and Teyssie, 1989).

Poly-β-hydroxy acid (e.g., poly-β-hydroxybutyrate (PHB) in Figure 1) can be prepared by both cationic and anionic ring opening polymerizations (Penczek and Kubisa, 1989; Penczek and Slomkowski, 1989, and Jerome and Teyssie, 1989). For example, 100% syndiotactic poly(β-DL-hydroxybutyrate) has been prepared by treating the corresponding lactone with cyclic dibutyltin initiators to yield high molecular weight polymers (Kricheldorf and Lee, 1995). Bacteria also produce the chiral, isotactic poly(β-D-hydroxybutyrate) as a highly crystalline biopolymer (Holmes, 1988). Numerous analogs of poly(β-hydroxy acid) have been synthesized either chemically by the ring opening polymerization or biologically by feeding unusual carbon sources to bacteria (Timmins and Lenz, 1994). The microbial synthesis of polyesters has been reviewed by Gross (1994). Due to their biocompatibility and biodegradability, different blends of polycaprolactone, poly(β-hydroxybutyrate) and other polymers have been fabricated for medical devices (Yasin and Tighe 1992), drug delivery applications (Wang, 1989), and cell microencapsulation (Giunchedi et al, 1994; Embleton and Tighe, 1993).

Surface-eroding polymer matrices may be attractive for a variety of tissue engineering applications. The monomer release would be steady over the lifetime of the matrices in contrast to PLA and PGA. In addition, the gradual loss of polymer from the surface of the scaffold may allow the surrounding tissue to serially fill the space vacated by the polymer. Polyorthoesters are an example of surface-eroding polymers. The hydrophobic character of the polymer limits water penetration and hydrolysis to only the exterior surface of the polymer matrix (Heller, 1985). Thus the surface erosion is much faster than that of the bulk. The chemical and physical properties of polyorthoesters have been reviewed (Heller and Daniels, 1994) and depend on the chemical structures of the constituent monomers. For example, reaction of bis(ketene acetal) with rigid *trans*-cyclohexane dimethanol produces a rigid polymer with a Tg of 110°C, whereas that of the flexible diol 1,6-hexanediol produces a soft material having a Tg of 20°C. Mixture of the two diol results in polymers having intermediate Tg. Degradation of the polymers is acid-induced, and degradation rates can be increased by adding acidic excipients or by increasing the hydrophilicity of the polymer matrix. Conversely, degradation can be retarded by using basic excipients such as $Mg(OH)_2$ (Gomboz and Pettit, 1995). Current applications of these polymers include sustained drug delivery as well as hard and soft tissue fixation (Heller and Daniels, 1994).

Polyorthoformate, polycarbonate, poly(oxyethylene glycolate), poly(1,4-butylene diglycolate), and polyurethane are other biodegradable polymers that may have applications in tissue engineering. Many of these polymers have been previously utilized as drug delivery matrices (Huang, 1989).

In summary, there is a large number of polyesters and analogs that are biodegradable. Their mechanical properties can be controlled largely by the chemical structures of the constituent building blocks and can be varied from tough to elastic. The biocompatibility of these polymers is assumed to result from nontoxic degradation products. There are efforts to attach bioactive elements to this class of materials in order to mimic natural extracellular matrix molecules.

Polypeptides

Proteins, one of the most important biomolecules in nature, belong to this class of biopolymers. However, polypeptides of a single amino acid or copolymers of two have long been regarded as impractical industrial materials (Nathan and Kohn, 1994). Amino acid N-carboxyanhydrides must be prepared as the monomeric starting materials, and this adds considerably to the cost of all polypeptides. These polymers are thus expensive even if they were derived from cheap amino acids. In addition, it is almost impossible to control the sequence of the protein polymers using random copolymerization techniques. Most polypeptides are insoluble in common organic solvents. The need for exotic solvents systems to process these materials combined with their thermal instability make them very poor engineering materials.

A number of recent approaches may, however, bypass these difficulties. Advances in genetic engineering have enabled investigators to obtain protein polymers by inserting DNA templates of predetermined sequences into the genome of bacteria. Collagenlike, silklike, and silk-elastinlike proteins have been synthesized by this technique (Cappello et al, 1990; Goldberg et al, 1989; McGrath et al, 1992). The general concept (reviewed by O'Brien, 1993) involves the incorporation of amino acid sequences with desired properties (e.g., cell adhesion or elasticity) into the protein polymers to produce materials of predetermined structure and controlled properties. For example, a cell adhering sequence of RGD has been incorporated into silklike protein polymers in a manner such that the tripeptide sequence is exposed for cell attachment (Tirrell et al, 1994). Investigators have also developed chemical synthetic techniques that are complementary to the genetic approach to prepare such materials. For instance, different rigid nonpeptide, organic segments have been combined with leucine-glutamine-proline, a sequence of the calcium binding domain of bovine amelogenins, using a completely synthetic approach (Sogah et al, 1994). The advantage of this class of protein-based hybrid polymers is the virtually unlimited choice of building blocks for the polymers. In contrast, genetically engineered proteins can only make use of the 20 natural and a limited number of unnatural amino acids for the construction of polymers. Rigid organic segments have been used to reduce the conformational flexibility of the peptide chain through the formation of peptide secondary structures (e.g., β-sheet or β-turns). Controlled folding of the polymer backbone has been reported using such ordered building blocks (Figure 5) (Wong, 1996). The potential application of these materials to tissue engineering is tremendous. These synthetic techniques allow precise control of material properties, while maintaining the freedom and flexibility to design proteinlike materials with desirable biological and chemical properties. These properties may make this class of materials superior matrices for tissue engineering applications.

Urry and coworkers have also studied elastin protein-based polymers as biocompatible materials. These polymers, also known as bioelastic materials, are elastomeric polypeptides comprised of the general repeating sequences glycine-any amino acid-glycine-valine-proline (GXGVP). The polymers were synthesized by the self-condensation of the activated p-nitrophenol ester of the pentapeptide building blocks (Prasad et al, 1985). The molecular weight of these polymers is considered to be higher than 50,000. The polymers can be cross-linked by γ-irradiation to form an insoluble matrix with-

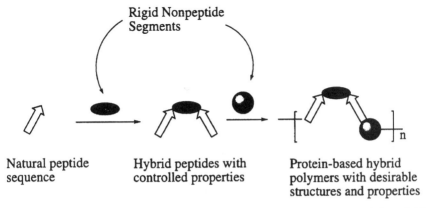

Natural peptide sequence

Hybrid peptides with controlled properties

Protein-based hybrid polymers with desirable structures and properties

Figure 5. Schematic drawing of protein-based hybrid polymers containing rigid nonpeptide segment that can be used to control the properties of the resulting polymers

out detectable residue destruction. Cell adhesion sequences (e.g., RGD) and enzymatic sites have been incorporated into the polymers for cell attachment and catalytic activity studies, respectively. Synthesis utilizing genetic engineering approach has also been reported. This class of polymers has been reported to exhibit excellent biocompatibility (Urry, 1993; Urry et al, 1995).

One specific polypeptide $(GVGVP)_n$ has been shown to undergo an inverse temperature transition in water (Urry, 1988a,b). The mechanism of such elasticity has been demonstrated to be entropic in nature and is apparently due to the internal chain dynamics of the ordered polypeptide structure. This is contrary to the common belief that the elasticity of elastin, similar to synthetic polymers, is due to random chain network and random end-to-end distances (Alberts et al, 1983). The transition temperature can be controlled by the amino acid composition, pH, and phosphorylation, electrochemical, photochemical, and chemical reactions of prosthetic groups. Therefore, a device that converts chemical energy into mechanical work can be constructed. Applications in surgery, tissue engineering, and drug delivery may be found (Urry et al, 1995).

Blends, interpenetrating networks (IPN) and composites

Although there are large number of polymer blends described in the literature, only those blends that contain biodegradable polymers and/or natural components are applicable under the context of tissue engineering. The use of polymer blends or composites (polymeric composite materials) as biomaterials is a concept that nature exploits in assembling ECM in tissue. The ECM of tissues typically contains a composite of different macromolecules and nonmacromolecular materials. For example, glycoaminoglycans, which are usually covalently linked to proteins to form proteoglycans, constitutes a gellike, highly hydrated structure substance in the which the collagen fibers are embedded (Giusti et al, 1993; Wight et al, 1991).

Blends of fibrin and polyurethane have previously been formed by a combined phase-inversion and spray process to produce highly porous small-diameter vascular prostheses (Giusti et al, 1985; Soldani et al, 1992). These materials exhibit high thermal stability, and their tensile behavior ranges from that of an elastic polyurethane tube to that of a natural blood vessel. Hydrogels of fibrin and poly(vinyl alcohol), blends or IPNs of collagen and poly(vinyl alcohol), blends of hyaluronic acid with poly(vinyl alcohol) or poly(acrylic acid), and blends based on esters of hyaluronic acid have been reported (Giusti et al, 1993). These materials may be suitable for a variety of applications including soft tissue replacement, drug delivery, nerve-guide growth, and cardiovascular devices. This class of materials has great potential owing to the large number of readily available synthetic polymers that can be mixed with biopolymers.

Polyanhydrides and poly(amino acids)

Both polyanhydrides and poly(amino acid)s have been utilized in drug delivery and tissue engineering studies. The syntheses and characterization of polyanhydrides and poly(amino acid)s as biodegradable polymers are reviewed elsewhere.

Immobilization materials

Immobilization materials are utilized to physically confine active biologic components (e.g., enzymes, proteins, cell organelles, and cells) while they are carrying out their biological functions. The choice of matrices for cell immobilization has been reviewed by Scouten (1995). We will consider those materials that are biodegradable and promising for tissue engineering applications.

Polysaccharides

Polysaccharides are carbohydrates characterized by the presence of a repeating structure in which the interunit linkages are of the O-glycoside type. The hydrophilicity of polysaccharides, along with the ease in which they can be formed into hydrogels, makes these materials ideal for many tissue engineering applications in which one desires to immobilize cells within a matrix. The variety of saccharides monomers ($\approx$200) and the variety of possible O-glycoside linkages result in a diversity of polysaccharide structures and conformations. Polysaccharides may be derived from different sources including plants (starch, cellulose), animals (glycogen), algae and seaweeds (alginate and agarose), and microorganisms. These materials are usually considered as naturally-derived products. However, since polysaccharides are widely utilized as immobilization materials, they are used here as standards to which other synthetic materials are compared.

Algal polysaccharides: alginate and agarose

Algal polysaccharides have been the most commonly utilized immobilization materials. This is due to their gentle gelling conditions, widespread availability, and relative biocompatibility. The main starting sources of alginate are species of brown algae (Phaeophycae). The algae are typically subjected to a number of processing steps to produce pure alginate which is the major polysaccharide present and may comprise up to 40% of the dry weight. It is part of the intracellular matrix and exists, in the native state, as a mixed salt of the various cations found in sea water (e.g., Mg^{2+}, Ca^{2+}, Sr^{2+}, Ba^{2+}, and Na^+). Due to selectivity of cation binding, the native alginate is mainly found in the insoluble gel form, which results from cross-linking of alginate chains by Ca^{2+}. All alginates are copolymers of D-mannuronate (M) and L-guluronate (G). However, alginates from different algal sources have different compositions, and thus, different physical and mechanical properties. The block length of monomer units and overall composition of the alginate and molecular weight determine the properties of alginates. For example, calcium alginates rich in G are stiff materials (Sutherland, 1991).

Alginate selectively binds divalent metal ions such as Ba^{2+}, Sr^{2+}, and Ca^{2+}. The binding selectivity increases with G content, and polyman-nuronate is essentially nonselective. The calcium ions are, therefore, selectively bound between sequences of polyguluronate residue and are held between diaxially linked L-guluronate residues which are in the $^1C_4$ chair conformation (Figure 6). The calcium ions are thus packed into the interstices between polyguluronate chains associated pairwise, and this structure is named the egg-box sequence (Figure 7). The ability to form a junction zone depends on the length of the G-blocks in different alginates. Since the mechanical strength of alginate gels depend on the block lengths and M/G

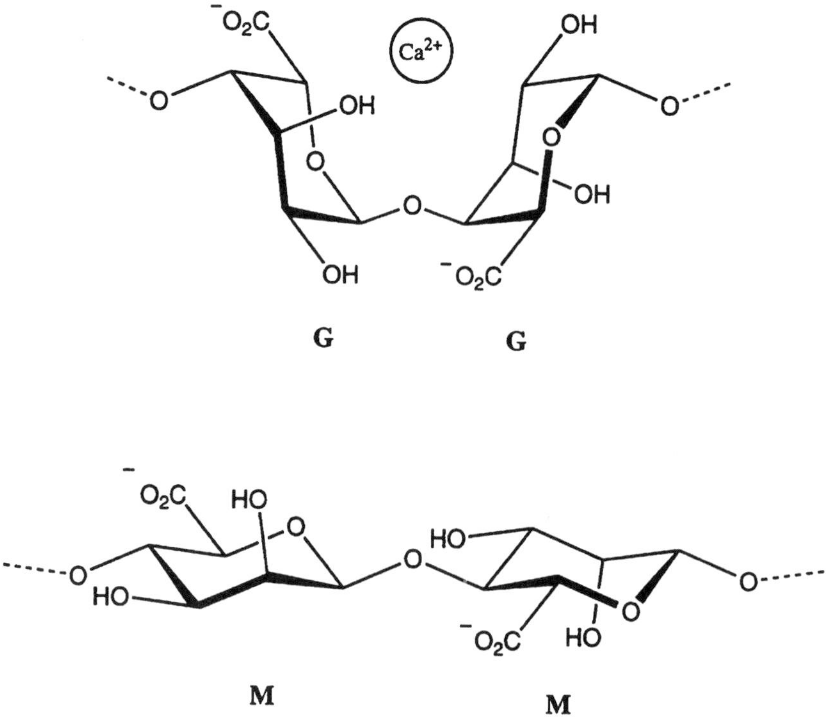

Figure 6. Structural geometry of MM and GG dimer in alginate and the calcium ion binding site for the GG dimer

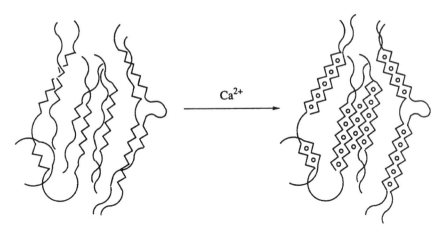

Figure 7. Schematic representation of a guluronate junction zone in alginate—the egg-box model. The circles represent calcium ions.

content, there have been efforts to modify the M/G ratio by alginase to increase the G content (Skjak-Braek et al, 1986). It is expected that chemically modified alginate would also produce materials of desirable properties. For example, bacterial alginates that contain acetyl groups generally exhibit different physical and mechanical properties from those of algal sources (Ott and Day, 1995).

Alginate can be gelled under mild conditions, allowing cell immobilization with little damage. Binding of Mg^{2+} and monovalent ions to alginate does not induce gelation of alginate in aqueous solution (Sutherland, 1991). However, exposure of alginate to soluble calcium leads to a preferential binding of calcium and subsequent gelling. These gentle gelling conditions are in contrast to the large temperature or solvent changes typically required to induce similar phase changes in most materials. Alginates have been utilized as immobilization matrices for cells (Smidsrod and Skjak-Braek, 1990), as an injectable matrix for engineering cartilaginous tissue to treat vesicoureteral reflux in various animal models (Atala et al, 1993; 1994), and as injectable microcapsules containing islet cells to treat animal models of diabetes (Sun et al, 1984). The open lattice structure and wide distribution of pore sizes in calcium alginate preclude the controlled release of large molecules (e.g., proteins) from these materials and limits the use of pure alginate for entrapment of whole cells or cell organelles (Smidsrod and Skjak-Braek, 1990). However, alginate membrane can be modified by incorporating other polymeric elements (e.g., lysine, poly(ethylene glycol), poly(vinyl alcohol), or chitosan) (Kung et al, 1995; Polk et al, 1994). These modified systems have been used to control the release of proteins from alginate beads. Haemostatic swabs made of calcium alginate have also been clinically utilized to reduce blood loss during surgical procedures. The calcium ions in alginate may assist the blood clotting process by activating platelets and clotting factor VII (Blair et al, 1990).

Agarose is another type of marine algal polysaccharide. In contrast to alginate, agarose forms thermally reversible gels. Agarose will set at concentrations in excess of 0.1%, depending on the sulfate content, and at temperatures considerably below ($\approx$40°C) the gel-melting temperature ($\approx$90°C). The latter parameter is correlated to the methoxy content. The proposed gel structure is bundles of associated double helices, and the junction zones consist of multiple chain aggregations (Yalpani, 1988). Agarose has been used largely in gels for electrophoresis of proteins and nucleic acids. However, agarose gels have also been used as supporting materials for electrophoresis of bacteriophages (Serwer, 1987) and migration studies of leukocytes (Kallen et al, 1977). Although applications in tissue engineering have not been reported, its adjustable gelling behavior may render low temperature melting agarose a suitable injectable and immobilization matrix material.

Other polysaccharides

Microbial polysaccharides are ubiquitous in nature and the most abundant biopolymers on earth. They are of interest because of their unusual and useful functional properties. Some of these properties are summarized by Kaplan et al (1994) as the following: (1) film-forming and gel-forming capabilities, (2) stability over broad temperature ranges, (3) biocompatibility (natural products avoid the release/leaching of toxic metals, residual chemicals, catalyst, or additives), (4) unusual rheological properties, (5) biodegradability, (6) water solubility in the native state or reduced solubility if chemically modified, and (7) thermal processability for some of these polymers. Some examples of microbial polysaccharide are listed in Table 5. It is worthy to note that gellan, one of the microbial polysaccharides, has been investigated as an immobilization material for enzymes and cells (Doner and Douds, 1995).

Nonnatural hydrogels

Polyphosphazenes

Polyphosphazenes contain inorganic backbones comprised of alternating single and double bonds between nitrogen and phosphorus atoms, in contrast to the carbon–carbon backbone in most other polymers. (Figure 1) The uniqueness of polyphosphazenes stems from the combination of this inorganic backbone with versatile side chain functionalities that can be tailored

Table 5. Some polysaccharides synthesized by microorganisms

Polymers[a]	Structure
Fungal	
Pullulan (N)	1,4-; 1,6-α-D-Glucan
Scleroglucan (N)	1,3-; 1.6-α-D-Glucan
Chitin (N)	1,4-β-D-Acetyl glucosamine
Chitosan (C)	1,4-β-D-N-Glucosamine
Elsinan (N)	1,4-; 1,3-α-D-Glucan
Bacterial	
Xanthan gum (A)	1,4-β-D-Glucan with D-mannose; D-glucuronic acid as side groups
Curdlan (N)	1,3-β-D-Glucan (with branching)
Dextran (N)	1.6-α-D-Glucan with some 1,2-; 1,3; 1,4-α-linkages
Gellan (A)	1,4-β-D-Glucan with rhamose, D-glucuronic acid
Levan (N)	2,6-β-D-Fructan with some β-2, 1-branching
Emulsan (A)	Lipoheteropolysaccharide
Cellulose (N)	1,4-β-D-Glucan

[a]N = neutral, A = anionic and C = cationic.
Adapted from Kaplan et al, 1994.

for different applications. The degradation of polyphosphazenes results in the release of phosphate and ammonium ions along with the side groups. The synthetic methods and medical applications of these polymers have previously been reviewed (Allcock, 1989; Scopelianos, 1994). Linear, un-cross-linked polymers can be prepared by thermal ring opening polymerization of $(NPCl_2)_3$ and the chloro group replaced by amines, alkoxides, or organometallic reagents to form hydrolytically stable, high molecular weight poly(organophosphazenes). Depending on the properties of the side groups, the polyphosphazenes can be hydrophobic, hydrophilic or amphiphilic. The polymers can be fabricated into films, membranes, and hydrogels for biomedical applications by cross-linking or grafting (Allcock et al, 1988a,b; Lora et al, 1991). Bioerodible polymers for drug delivery devices have been prepared by incorporating hydrolytic side chains of imidazole for skeletal tissue regeneration (Laurencin et al, 1993). Nondegradable phosphazenes have been used as denture liner (Razavi et al, 1993).

Poly(vinyl alcohol) (PVA)

PVA (Figure 1) is not synthesized directly but is the deacetylated product of poly(vinyl acetate). Polyvinyl acetate is usually prepared by radical polymerization of vinyl acetate (bulk, solution, or emulsion polymerizations) (Finch, 1973). PVA is formed by either alcoholysis, hydrolysis, or aminolysis processes of poly(vinyl acetate). The hydrophilicity and water solubility of PVA can be readily controlled by the extent of hydrolysis and molecular weight. PVA has been widely used as a thickening and wetting agent. PVA gels can be prepared by cross-linking with formaldehyde in the presence of sulfuric acid (Schwartz and Erich, 1960). These formaldehyde-cross-linked PVA materials have been used as prostheses for a variety of plastic surgery applications including breast augmentation (Clarkson, 1960; Peters and Smith, 1981), diaphragm replacement (Haupt and Myers, 1960), and bone replacement (Camerson and Lawson, 1960). However, a variety of complications were found after long term implantation, including calcification of the PVA (Peters and Smith, 1981). More recently, PVA was made into an insoluble gel using a physical cross-linking process. These gels were prepared with a repeated freezing-thawing process. This causes structural densification of the hydrogel due to the formation of semicrystalline structures. The use of this gel in drug delivery applications has been reported (Ficek and Peppas, 1993; Peppas and Scott, 1992). However, PVA is not truly biodegradable due to the lack of labile bonds within the polymer bond. Only low molecular weight materials are advisable to be used as implant materials.

Poly(ethylene oxide) (PEO)

PEO or polyethylene glycol (Figure 1) can be produced by the anionic or cationic polymerization of ethylene oxide using a variety of initiators (Boileau, 1989; Penczek and Kubisa, 1989). PEO is highly hydrophilic and biocompatible, and has been utilized in a variety of biomedical applications including preparation of biologically relevant conjugates (Zalipsky, 1995), induction of cell membrane fusion (Lentz, 1994), and surface modification of biomaterials (Amiji and Park, 1993). Different polymer architectures have been synthesized, and some of their applications in medicine have been recently reviewed (Merrill, 1993). For example, PEO can be made into hydrogels by γ-ray or electron beam irradiation and chemical cross-linking (Belcheva et al, 1996; Cima et al, 1995). These hydrogels have been used as matrices for drug delivery and cell adhesion studies.

Pluronics

Pluronic polyols or polyoxamers are block copolymers of PEO and poly(propylene oxide) (Figure 1) and are usually synthesized by anionic polymerization in the form of a ABA triblock using a difunctional initiator (Schmolka, 1972). Pluronics F127, which contains 70% ethylene oxide and 30% propylene oxide by weight with an average molecular weight of 11,500, is the most commonly used gel-forming polymer matrix to deliver proteins (Gombotz and Pettit, 1995). This polymer exhibits a reversible thermal gelation in aqueous solutions at a concentration of 20% or more (Schmolka,

1972). Thus, the polymer solution is a liquid at room temperature but gels rapidly in the body. Although the polymer is not degraded by the body, the gels dissolve slowly and the polymer is eventually cleared. This polymer has been utilized in protein delivery (Jushasz et al, 1989; Morikawa et al, 1987) and skin burn treatments (Pautian et al, 1993).

PGA-PEO hydrogels

Although PGA is not water soluble, bioerodible hydrogels based on pho-topolymerized PGA-PEO copolymers have been synthesized and their biological activities investigated (Hill-West et al, 1994; Sawhney et al, 1993; 1994). Macromonomers having a poly(ethylene glycol) central block, extended with oligomers of α-hydroxy acids (e.g., oligo(dl-lactic acid) or oligo(glycolic acid)) and terminated with acrylate groups were synthesized. These hydrogels were designed to form direct contacts with tissues or proteins following photopolymerization and act as a barrier. These gels degrade upon hydrolysis of the oligo(α-hydroxy acid) regions into poly(ethylene glycol), the α-hydroxy acid, and oligo(acrylic acid). The degradation rate of these gels could be tailored from less than 1 day to 4 months by appropriate choice of the oligo(α-hydroxy acid). The macromonomer could be polymerized using nontoxic photoinitiators with visible light without excess heating or local toxicity. The hydrogels polymerized in contact with tissue adhere tightly to the underlying tissue. In contrast, the gels were nonadhesive if they were polymerized prior to contact with tissue. These hydrogels have been utilized in animal models to prevent post-surgical adhesion and thrombosis of blood vessels and initimal thickening following ballon catheterization.

CONCLUSION

How matrices can be utilized in tissue engineering

Biodegradable synthetic polymers show tremendous potential as matrices to engineer new tissues. For example, Figure 8 shows a SEM photograph of cultured chrondocytes on a PGA scaffold. The fibrous polymer matrix provides large pores for the chrondocytes to penetrate, and it also provides a three-dimensional physical space and mechanical support for the attached cells as they proliferate and assemble a cartilaginous extracellular matrix on the polymer. This process ultimately results in the formation of new cartilage, and the scaffold can be used to control the structure of this tissue *in vitro* and *in vivo* (Ma et al, 1995; Puelacher et al, 1994).

Tissue engineering is an interdisciplinary field in which cell and molecular biology, physiology, immunology, biological chemistry, biophysics, and sur-

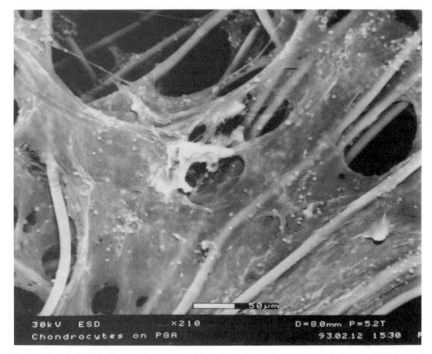

Figure 8. A SEM photomicrograph showing chrondocytes seeded on a PGA scaffold. (Mooney DJ, unpublished results). The rods are the PGA fiber, the dots are the cultured chrondocytes, and spanning the fibers are extracellular matrix proteins and glycosaminoglycans secreted by the cells. (Freed and Vunjak-Novakovic, 1995)

gery are involved in addition to the polymer and materials science issues discussed in this chapter. Applications of biodegradable polymers in tissue engineering are demonstrated in other chapters of this book, which also address various elements of these other issues.

Outlook

The vast number of synthetic biodegradable polymers offers tremendous opportunities to tissue engineering. Established polymer chemistries enable one to easily tailor properties of the synthetic polymers by using different (1) functional groups (either on the backbone or side chain), (2) polymer architectures (linear, branched, comb or star), and (3) combinations of polymer species physically mixed (polymer blends or interpenetrating networks) or chemically bonded (copolymers). Biomaterials scientists have harnessed this diversity of synthetic polymers to fabricate a number of devices for different tissue replacements (e.g., silicon rubber and telfon as catheters and vascular grafts) (Ratner, 1992). In addition, drug delivery applications have encouraged studies of degradation, physical, and mechanical properties of many biodegradable polymers (Gombotz and Pettit, 1995). However, α-hydroxy polyesters and, to a lesser extent, alginates have dominated the new field of tissue engineering. The preference for PGA and related polyesters is partially due to their established safety in human applications and the projected approval of the Food and Drug Administration. Various properties of PGA/PLA make them good tissue engineering matrices, but they are far from the perfect ones. For example, PGA/PLA are stiff materials which make them unsuitable matrices for soft tissue engineering. The bulk degradation mechanism of these polyesters causes very rapid loss of mechanical strength after implantation. In addition, PLGA are largely inert to the body unless specific peptide sequences have been incorporated into the polymer. Polymers constituted of building blocks similar to components of ECM (e.g., carbohydrates and peptides) may be more appropriate in some cases. This chapter is written to encourage investigators to appreciate other polymer systems along with the PLGA family. A large amount of previous research in related fields may ease the adaptation of these polymers to tissue engineering applications. In addition, the matrices utilized to engineer different tissues may have very diverse requirements. Designing polymer systems to tailor their biological and mechanical properties for certain tissue engineering applications may be a better approach than simple adaptation. Living systems are very complex, and engineering tissue is a challenging and complicated task that is worthy of a significant intellectual investment.

Acknowledgments The authors thank NSF (grant #BES-9501376), the Whitaker Foundation and Reprogenesis for financial support.

References

Alberts B, Bray D, Lewis J, Raff M, Roberts K, Watson JD (1983): *Molecular Biology of The Cell.* New York: Garland Publishing

Allcock HR (1989): Phosphazene high polymers. In: *Comprehensive Polymer Science,* Vol. 4, Allen G, ed. New York: Pergamon Press

Allcock HR, Gebura M, Kwon S, Neenan TX (1988a): Amphiphilic polyphosphazenes as membrane materials: Influence of side group on radiation cross-linking. *Biomaterials* 9:500–508

Allcock HR, Kwon S, Riding GH, Fizzpatrick RJ, Bennett JL (1988b): Hydrophilic polyphosphazenes as hydrogels: radiation cross-linking and hydrogel characteristics of poly[bis(methyethoxythoxy)-phosphazene. *Biomaterials* 9:509–513

American Liver Foundation (1988): *Vital statistics of the United States,* vol. 2, part A. New Jersey: ALF

Amiji M, Park K (1993): Surface modification of polymeric biomaterials with poly(ethylene oxide), albumin, and heparin for reduced thrombogenicity. *J Biomater Sci Polym Ed* 4:217–234

Atala A, Cima LG, Kim W, Paige KT, Vacanti JP, Retik AB, Vacanti CA (1993): Injectable alginate seeded with chrondrocytes as a potential treatment for vesicoureteral reflux. *J Urology* 150:745–747

Atala A, Kim W, Paige KT, Vacanti CA, Retik AB (1994): Endosopic Treatment of vescoureteral reflux with a chrondrocyte-alginate suspension. *J Urology* 152:641–643

Barakat I, Dubois P, Grandfils C, Jerome R (1996): Macromolecular engineering of polylactone and polylactide. XXI. Controlled synthesis of low molecular weight polylactide macromonomers. *J Polym Sci Polym Chem* 34:497–502

Barrera DA, Zylstra E, Lansbury PT, Langer R (1993): Synthesis and RGD peptide modification of a new biodegradable copolymers: Poly(lactic acid-*co*-lysine). *J Am Chem Soc* 115:11010–11011

Belcheva N, Stamenova R, Tsvetanov C (1996): Crosslinking poly(ethylene oxide) for drug release systems. *Macromol Symp* 103:193–211

Blair SD, Jarvis P, Salmon M, McCollum C (1990): Clinical trial of calcium alginate haemestatic swabs. *Br J Surg* 77:568–570

Boileau S (1989): Anionic ring-opening polymerization: Epoxides and episulfides. In: *Comprehensive Polymer Science,* Vol. 3, Allen G, ed. New York: Pergamon Press

Brittberg M, Lindahl A, Nilsson A, Ohlsson C, Isaksson O, Peterson L (1994): Treatment of deep cartilage defects in the knee with autologous chrondrocyte transplantation. *NEJM* 331:889–895

Camerson JM, Lawson DD (1960): The failure of polyvinyl sponge as a bone substitute. *Res Vet Sci* 1:230–231

Cappello J, Crissman J, Dorman M, Mikolajckak M, Textor G, Marquet M, Ferrari F (1990): Genetic engineering of structural protein polymers. *Biotechnol Prog* 6:198–202

Chujo K, Kobayashi H, Suzuki J, Tokuhara S (1967): Physical and chemical characteristics of polyglycolide. *Makromol Chem* 100:267–270

Chukwuma C Sr (1995): Type II diabetic nephropathy in perspective. *J Diabetes Complications* 9:55–67

Cima LG, Lopina ST, Merrill EW (1995): Hepatocyte response to PEO-tethered carbohydrates depend on tether conformation. *Transactions of the 21st Annual meeting of the Society for Biomaterials* 18:147

Clarkson P (1960): Sponge implants for flat breasts. *Proc R Soc of Med* 53:880–881

Cooper ML, Hansbrough JF, Speilvogel RL, Cohen R, Bartel RL, Naughton G (1991): In vivo optimization of a living dermal substitute employing cultured human fibroblasts on a biodegradable polyglycolic acid or polyglactin mesh. *Biomaterials* 12:243–248

Doner LW, Douds DD (1995): Purification of commercial gellan to monovalent cation salts results in acute modification of solution and gel-forming properties. *Carbohydr Res* 273:225–233

Embleton JK, Tighe BJ (1993): Polymers for biodegradable medical devices. X. Microencapsulation studies: Control of poly-hydroxybutyrate-hydrovalerate microcapsules porosity via polycaprolactone blending. *J Microencapsul* 10:341–352

Engelberg I, Kohn J (1991): Physico-mechanical properties of degradable polymers used in medical applications: A comparative study. *Biomaterials* 12:292–304

Ficek BJ, Peppas NA (1993): Novel preparation of poly(vinyl alcohol) microparticles without crosslinking agent for controlled drug delivery of proteins. *J Controlled Release* 27:259–264

Finch CA (1973): *Polyvinyl Alcohol: Properties and Applications.* London: Wiley

Frazza EJ, Schmitt EE (1971): A new absorbable suture. *J Biomed Mater Res Symp* 1:43–58

Freed LE, Vunjak-Novakovic G (1995): Tissue engineering of cartilage. In: *The Biomedical Engineering Handbook,* Bronzino JD, ed. Boca Raton: CRC Press

Gilding DK, Reed AM (1979): Biodegradable polymers for use in surgery - polyglycolic/poly(lactic acid) homo- and copolymers: 1. *Polymer* 20:1459–1464

Giunchedi P, Conti B, Maggi L, Conte U (1994): Cellulose acetate butyrate and polycaprolactone for ketoprofen spray-dried microsphere preparation. *J Microencapsul* 11:381–393

Giusti P, Soldani G, Palla M, Paci M, Levita G (1985): New biolized polymers for cardiovascular applications. *Life Support Syst* 3(Suppl 1):476–480

Giusti P, Lazzeri L, Lelli L (1993): Bioartificial polymeric materials: A new method to design biomaterials by using both biological and synthetic polymers. *Tr Polym Sci* 9:261–266

Goldberg I, Salerno AJ, Patterson T, Williams JI (1989): Cloning and expression of a collagen-analog-encoding synthetic gene in *Escherichia coli. Gene* 80:305–314

Gombotz WR, Pettit DK (1995): Biodegradable polymers for protein and peptide drug delivery. *Bioconjugate Chem* 6:332–351

Gross RA (1994): Bacterials polyesters: Structural variability in a microbial synthesis. In: *Biomedical Polymers: Designed-to-Degrade Systems,* Shalaby SW, ed. Munich: Carl Hanser Verlag

Gupta MC, Desmuth VG (1983): Radiation effect on poly(lactic acid). *Polymer* 24:827–830

Hansbrough JF, Cooper ML, Cohen R, Spielvogel R, Greenleaf G, Nartel RL, Naughton G (1992): Evaluation of a biodegradable matrix containing cultured human fibroblasts as a dermal replacement beneath meshed skin graft on athymic mice. *Surgery* 111:438–446

Haupt GJ, Myers RD (1960): Polyvinyl formalinized (Ivalon) sponge in the repair of diaphragmatic defects. *A M A Arch Surg* 80:613–615

Hay DL, von Fraunhofe JA, Chegini N, Masterson BJ (1988): Locking mechanism strength of absorbable ligating devices. *J Biomed Mater Res* 22:179–190

Heller J (1985): Controlled drug release from poly(orthoester)-a surface eroding polymer. *J Controlled Release* 2:167–177

Heller J, Daniels AU (1994): Poly(ortho esters). In: *Biomedical Polymers: Designed-to-Degrade Systems,* Shalaby SW, ed. Munich: Carl Hanser Verlag

Hill-West JL, Chowdhury SM, Slepian MJ, Hubbell JA (1994): Inhibition of thrombosis and intimal thickening by *in situ* photopolymerization of thin hydrogel barrier. *Proc Natl Acad Sci USA* 91:5967–5971

Hoelscher DD, Weir MR, Bakris GL (1995): Hypertension in diabetic patients: An update of interventional studies to preserve renal function. *J Clin Pharmacol* 35:73–80

Holmes PA (1988): Biologically produced (R)-3-hydroxyalkane polymers and copolymers. In: *Developments in Crystalline Polymers,* Bassett DC, ed. New York: Elsevier

Huang SJ (1989): Biodegradable Polymers. In: *Polymers—Biomaterials and Medical Applications,* Kroschwitz JI, ed. New York: Wiley & Sons

Hubbell JA (1995): Biomaterials in tissue engineering. *Bio/Technology* 13:565–576

Jerome R, Teyssie P (1989): Anionic ring-opening polymerization. In: *Comp Polym Sci,* Vol. 3. New York: Pergamon Press

Johnson LB, Aiken J, Mooney D, Schloo BL, Griffith-Cima L, Langer R, Vacanti JP (1994): The mesentery as a laminated vascular bed for hepatocyte transplantation. *Cell Transplantation* 3:273–281

Jushasz J, Lenaerts V, Raymond P, Ong H (1989): Diffusion of rat atrial natriuretic factor in thermoreversible poluoxamer gels. *Biomaterials* 10:265–268

Kallen B, Nilsson O, Thelin C (1977): Effect of encephalitogenic protein on migration in agarose of leukocytes from patients with multiple sclerosis. Variable effect of the antigen in a large dose range, with a literature review. *Acta Neurol Scand* 55:33–46

Kaplan DL, Wiley BJ, Mayer JM, Arcidiacono S, Keith J, Lombardi SJ, Ball D, Allen AL (1994): Biosynthetic polysaccharides. *In: Biomedical Polymers: Designed-to-Degrade Systems.* Munich: Carl Hanser Verlag

Kasperczky J (1996): Microstructural analysis of poly[(L,L-lactide)-*co*-(glycolide)] by ^{1}H and [13]C n.m.r. spectroscopy. *Polymer* 37:201–203

Katz AR, Turner R (1970): Evaluation of tensile and absorption properties of poly-glycolic acid sutures. *Surg Gynecol Obstet* 131:701–716

Kim TH, Vacanti JP (1995): Tissue engineering of the liver. *In: The Biomedical Engineering Handbook.* Boca Raton: CRC Press

Kricheldorf HR, Kreiser-Saunders I (1996): Polylactides—synthesis, characterization and medical applications. *Macromol Symp* 103:85–102

Kricheldorf HR, Lee SR (1995): Polylactones. 35. Macrocyclic and stereoselective polymerization of β-DL-butyrolactone with cyclic dibutyltin initiators. *Macromol* 28:6718–6725

Kung IM, Wang FF, Chang YC, Wang YJ (1995): Surface modifications of algi-nate/poly(L-lysine) microcapsular membrane with poly(ethylene glycol) and poly(vinyl alcohol). *Biomaterials* 16:649–655

Langer R, Vacanti JP (1993): Tissue engineering. *Science* 260:920–926

Laurencin CT, Norman ME, Elgendy HM, el-Almin SF, Allcock HR, Pucher SR, Ambrosio AA (1993): Use of polyphosphazenes for skeletal tissue regeneration. *J Biomed Mater Res* 27:963–973

Leenslag JW, Pennings AJ, Bos RRM, Rozema FR, Boering G (1987): Resorbable materials of poly(L-lactide). VII. In vivo and in vitro degradation. *Biomaterials* 8:311–314

Lentz BR (1994): Polymer-induced membrane fusion: Potential mechanism and relation to cell fusion events. *Chem Phys Lipids* 73:91–106

Levesque L, Brubaker PL, Sun AM (1992): Maintenance of long-term secretary function by microencapsulated islets of Langerhans. *Endrocrinology* 130:644–650

Lora S, Carenza M, Palma G, Caliceti P, Battaglia P, Lora A (1991): Biocompatible polyphosphazene by radiation-induced graft copolymerization and hepariniza-tion. *Biomaterials* 12:275–280

Ma PX, Schloo B, Mooney D, Langer R (1995): Development of biomechanical properties and morphogenesis of in vitro tissue engineered cartilage. *J Biomed Mater Res* 29:1587–1595

Massia SP, Hubbell JA (1991): An RGD spacing of 440 nm is sufficient for integrin alpha V beta 3-mediated fibroblast spreading and 140 nm for focal contact and stress fiber formation. *J Cell Biol* 114:1089–1100.

Matas AJ, Sutherland DE, Steffes MW, Mauer SM, Sowe A, Simmons RL, Najarian JS (1976): Hepatocellular transplantation for metabolic deficiencies: Decrease of plasma bilirubin in Gunn rats. *Science* 192:892–4

McGrath KP, Fournier MJ, Mason TL, Tirrell DA (1992): Genetically directed synthesis of new polymeric materials. Expression of artificial genes encoding proteins with repeating -(AlaGly)$_3$ProGluGly- elements. *J Am Chem Soc* 114:727–733

Merrell JC, Russell RC, Zook EG (1986): Polyglycolic acid tubing as a conduit for nerve regeneration. *Ann Plast Surg* 17:49–58

Merrill EW (1993): Poly(ethylene oxide) star molecules: Synthesis, characterization, and applications in medicine and biology. *J Biomater Sci, Polym Ed* 5:1–11

Mooney DJ, Langer RS (1995): Engineering biomaterials for tissue engineering. The 10–100 micron size scale. In: *The Biomedical Engineering Handbook,* Bronzino JD, ed. Boca Raton: CRC Press

Mooney DJ, Breuer MD, McNamara K, Vacanti JP, Langer R (1995a): Fabricating tubular devices from polymers of lactic ad glycolic acid for tissue engineering. *Tissue Eng* 1:107–118

Mooney DJ, Park S, Kaufmann PM, Sano K, McNamara K, Vacanti JP, Langer R (1995b): Biodegradable sponges for hepatocyte transplantation. *J Biomed Mater Res* 29:959–965

Mooney DJ, Kaufmann PM, Sano K, Schwendeman SP, Majahod K, Schloo B, Vacanti JP, Langer R (1996): Localized delievery of epidermal growth factor improve the survival of transplanted hepatocytes. *Biotech Bioeng* 49:1–8

Morikawa K, Okada O, Hosokawa M, Kobayashi H (1987): Enhancement of thera-peutic effects of recombinant interleukin-2 on a transplantable rat fibrosarcoma by the use of a sustained release vehicle, pluronic gel. *Cancer* 47:37–41

Nakamura T, Hitomi S, Watanabe S, Shimizu Y, Jamshidi K, Hyon SH, Ikada Y (1989): Bioabsorption of polylactides with different molecular properties. *J Biomed Mater Res* 23:1115–1130

Nakamura T, Shimizu Y, Watanabe S, Hitomi S, Kitano M, Tamada J, Matsunobe S (1990): New bioabsorbable plegets and non-woven fabrics made from polygly-colide (PGA) for pulmonary surgery: Clinical experience. *Thorac Cardiovas Surg* 38:81–85

Nathan A, Kohn J (1994): Amino acid derived polymers. In: *Biomedical Polymers: Designed-to-Degrade Systems,* Shalaby SW, ed. Munich: Carl Hanser Verlag

Nerem RM, Sambanis A (1995): Tissue engineering: From biology to biological substitutes. *Tissue Eng* 1:3–13

O'Brien JP (1993): New structural polymers from Nature. *Tr Polym Sci* 8:228–232

Ott CM, Day DF (1995): Bacterials alginates: An alternative industrial polymer. *Tr Polym Sci* 3:402–406

Pautian PW, McPherson JC, Haase RR, Runner RR, Plowman KM, Ward DF, Nuyen TH, McPherson JC (1993): Intravenous Pluronic F-127 in early burn wound treat-ment in rats. *Burns* 19:187–191

Penczek S, Kubisa P (1989): Cationic ring-opening polymerization: Ethers. In: *Comprehensive Polymer Science,* Vol. 3, Allen G, ed. New York: Pergamon Press

Penczek S, Slomkowski S (1989): Cationic ring-opening polymerization: Cyclic ester. In: *Comprehensive Polymer Science,* Vol. 3, Allen G, ed. New York: Pergamon Press

Peppas NA, Scott JE (1992): Controlled release from poly(vinyl alcohol) gels prepared by freezing-thrawing processes. *J Controlled Release* 18:95–100

Peters WJ, Smith DC (1981): Ivalon breast prostheses: Evaluation 19 years after implantation. *Plast Rescontr Surg* 67:514–518

Pham HN, Padilla JA, Nguyen KD, Rosen JM (1991): Comparison of nerve repair techniques: Suture vs. avitene-polyglycolic acid tube. *J Reconstr Microsurg* 1:31–36

Polk A, Amsden B, De Yao K, Peng T, Goosen MF (1994): Controlled release of albumin from chitosan-alginate microcapsules. *J Pharm Sci* 83:178–185

Ponder KP, Gupta S, Leland F, Darlington G, Finegold M, DeMayo J, Ledley FD, Chowdhury JR, Woo SL (1991): Mouse hepatocytes migrate to liver parenchyma and function indefinitely after intrasplenic transplantation. *Proc Natl Acad Sci USA* 88:1217–1221

Prasad KU, Iqbal MA, Urry DW (1985): Utilization of 1-hydroxybenzotriazole in mixed anhydride coupling reactions. *Int J Pept Protein Res* 25:408–13

Puelacher WC, Mooney D, Langer R, Upton J, Vacanti JP, Vacanti CA (1994): Design of nasoseptal cartilage replacements synthesized from biodegradable polymers and chrondrocytes. *Biomaterials* 15:774–778

Ratner BD (1992): New ideas in biomaterials science—a path to engineered biomaterials. *J Biomed Mater Res* 27:837–850

Razavi R, Khan Z, Haenerle CB, Beam D (1993): Clinical applications of a polyphosphazene-based resilent denture liner. *J Prosthodontics* 2:224–227

Reed AM, Gilding DK (1981): Biodegradable polymers for use in surgery—poly(glycolic)/poly(lactic acid) homo and copolymers: 2. In vitro degradation. *Polymer* 22:494–498

Sanders RJ, Wilson MR (1993): Diabetes-related eye disorders. *J Natl Med Assoc* 85:104–8

Sawhney AS, Pathak CP, Hubbell JA (1993): Bioerodible hydrogels based on photopolymerized poly(ethylene glycol)-co-poly(α-hydroxy acid) diacrylate macromonomers. *Macromol* 26:581–587

Sawhney AS, Pathak CP, van Rensburg JJ, Dunn RC, Hubbell JA (1994): Optimization of photopolymerized bioerodible hydrogel properties for adhesion prevention. *J Biomed Mater Res* 28:831–838

Schmolka IR (1972): Artificial skin. I. Preparation and properties of pluronic F-127 gels for treatments of burns. *J Biomed Mater Res* 6:571–582

Schwartz AW, Erich JB (1960): Experimental study of polyvinyl-alcohol-formal (Ivalon) sponge as a substitute for tissue. *Plast Rescontr Surg* 25:1

Scopelianos AG (1994): Polyphophazenes as new biomaterials. In: *Biomedical Polymers: Designed-to-Degrade Systems,* Shalaby SW, ed. New York: Carl Hanser Verlag

Scouten WH (1995): Matrices and immobilization methods for cell adhesion/immobilization studies. *Bioprocess Technol* 20:233–265

Serwer P (1987): Agarose gel electrophoresis of bacteriophages and related particles. *J Chromatogr* 418:345–357

Shalaby SW, Johnson RA (1994): Synthetic absorbable polyesters. In: *Biomedical Polymers,* Shalaby SW, ed. Munich: Carl Hanser Verlag

Skjak-Braek G, Smidsrod O, Larsen B (1986): Tailoring of alginates by enzymatic modification in vitro. *Int J Biol Macromol* 8:330–336

Smidsrod O, Skjak-Braek G (1990): Alginate as immobilization matrix for cells. *Tr Biotech* 8:71–78

Sogah DY, Perl-Treves D, Wong WH, Zheng QY (1994): De Novo design and synthesis of protein-based hybrid polymers. *Macromol Reports* A31:1003–1008

Soldani G, Panol G, Sasken HF, Goddard MB, Galletti PM (1992): Small diameter polyurethane-polydimethylsiloxane vascular prostheses made by a spraying, phase-inversion process. *J Mater Sci Mater Med* 3:106–113

Sun AM, O'Shea GM, Goosen MF (1984): Injectable microencapsulated islet cells as a bioartificial pancreas. *Appl Biochem Biotech* 10:87–99

Sutherland IW (1991): Alginates. In: *Biomaterials: Novel Materials From Biological Sources,* Byrom D, ed. New York: Stockton Press

Thomson RC, Yaszemski MJ, Powers JM, Mikos AG (1995): Fabrication of biodegradation polymer scaffolds to engineer trabecular bone. *J Biomater Sci Polym Edn* 7:23–38

Timmins MR, Lenz RW (1994): Enzymatic biodegradation of polymers: The polymer chemist's perspective. *Tr Polymer Sci* 2:15–19

Tirrell JG, Fournier MJ, Mason TL, Tirrell DA (1994): Biomolecular Materials. In: *C&EN* (December 19), Jacobs M, ed. Washington, DC: American Chemical Society

Urry DW (1988a): Entropic elastic processes in protein mechanisms. I. Elastic structure due to an inverse temperature transition and elasticity due to internal chain dynamics. *J Protein Chem* 7:1–34

Urry DW (1988b): Entropic elastic processes in protein mechanisms. II. Simple (passive) and coupled (active) development of elastic forces. *J Protein Chem* 7:81–114

Urry DW (1993): Molecular machines: How motion and other function of living organisms can result from reversible chemical changes. *Angew Chem Int Ed Engl* 32:819–841

Urry DW, Parker TM, Minehan DS, Nicol A, Pattanaik A, Peng SQ, McPherson DT, Gowda DC (1995): The capacity to vary the bioactive elastic protein-based polymers. *PMSE Prepr, August 1995* 399–402.

Vacanti JP, Morse MA, Saltzman WM, Domb AJ, Perez-Atayde A, Langer R (1988): Selective cell transplantation using bioabsorbable artificial polymers as matrices. *J Ped Surg* 23:3–9

Vainionpaa S, Kilpikari J, Laiho J, Helevirta P, Rokkanen P, Tormala P (1987): Strength and strength retention in vitro, of absorbable, self-reinforced polyglycolide (PGA) rods for fracture fixation. *Biomaterials* 8:46–48

Vert M, Guerin P (1991): Biodegradable aliphatic polyesters of the poly(hydroxy acid)-type for temporary therapeutic applications. In: *Biomaterials Degradation,* Barbosa MA, ed. New York: Elsvier

Vert M, Christel P, Chalot F, Leray J (1984): Bioresorbable plastic materials for bone surgery. In: *Macromolecular Biomaterials,* Hastings GW, Ducheyne P, eds. Boca Raton: CRC Press

Wang PY (1989): Compressed poly(vinyl alcohol)-polycaprolactone admixture as a model to evaluate erodible implants for sustained drug delivery. *J Biomed Mater Res* 23:91–104

Wong WH (1996): *Design, Synthesis and Properties of Protein Based Hybrid Polymers.* PhD Thesis. Ithaca, NY: Cornell University.

Wight TN, Heinegrad DK, Hascall VC (1991): Proteoglycans: Structure and Function. In: *Cell Biology of Extracellular Matrix,* Hay ED, ed. New York: Plenum Press

Yalpani M (1988): *Polysaccharides: Syntheses, Modifications and Structure/Properties Relations.* Amsterdam: Elsevier

Yasin M, Tighe BJ (1992): Polymers for biodegradable medical devices. VIII. Hydroxybutyrate-hydroxyvalerate copolymers: Physical and degradative properties of blends with polycaprolactone. *Biomaterials* 13:9–16

Yerdel MA, Koksoy C, Karayalcin K, Kuzu A, Aoyama M, Anadol E (1993): Use of polyglycolic acid mesh with a new double pursestring technique in renal trauma: An experimental study. *Injury* 24:158–160

Zalipsky S (1995): Functionalized poly(ethylene glycol) for preparation of biologically relevant conjugates. *Bioconjugate Chem* 6:150–165

5

BIOACTIVE POLYMERS

Jennifer L. West and Jeffrey A. Hubbell

Introduction

Tissue engineering, which includes the development of replacements for damaged or diseased tissues and organs, is a rapidly growing new field that draws upon the expertise of chemists, chemical engineers, and materials scientists in addition to biologists and physicians. Polymeric biomaterials have played an important role thus far in tissue engineering and hold great promise for future applications. Some of the uses of polymeric biomaterials in tissue engineering have included immunoprotective membranes for cell transplantation (Pathak et al, 1992; Sefton et al, 1987) and scaffolds to support and guide tissue growth (Cima et al, 1990; Freed et al, 1994; Mooney et al, 1995).

Bioactive polymers are materials based on synthetic polymers with the additional incorporation of a bioactive species, such as proteins, peptides, or carbohydrates, into the polymer structure to induce a specific biological response. Such bioactive species may be immobilized or releasable, depending on the mechanism of action of the agent. Much of the current research in bioactive polymers has focused on polymers that support biospecific cell adhesion, e.g., adhesion to an oligopeptide via a cell surface receptor that specifically recognizes and binds to the amino acid sequence of the oligopeptide. While many cell adhesion peptides recognize receptors on a broad variety of cell types, some peptide sequences may be cell-type selective, thus allowing the possibility to create materials that are adhesive towards one type of cell but not to others. It may be possible to achieve a degree of cell-type specific adhesion based on cell-surface receptor-mediated interactions with carbohydrates. In addition, bioactive materials may

Synthetic Biodegradable Polymer Scaffolds
Anthony Atala and David Mooney, Editors
Robert Langer & Joseph P. Vacanti, Associate Editors
© 1997 Birkhäuser Boston

also contain species that alter cell function through a pharmacological mode of action, e.g., by blocking specific receptor or enzyme pathways with releasable compounds. These various approaches will be illustrated in the pages below.

CELL NONADHESIVE POLYMERS

A major goal in bioactive polymer research for tissue engineering is the development of materials that are capable of interacting specifically with certain cell types, while not allowing adhesion of other cell types. Cell nonadhesive polymers are briefly discussed here because if cellular adhesion to a bioactive material is to occur in a biospecific fashion, the base polymer must not support cell adhesion on its own.

Cell adhesion is normally mediated by the interaction of a cell adhesion protein (e.g., fibronectin) or proteoglycan with a cell-surface receptor (e.g., an integrin). Clearly, cells do not possess receptors for synthetic polymers, so one might expect that synthetic polymers would not support attachment of cells. However, when polymeric materials are implanted in vivo or seeded with cells in vitro, virtually all readily support cell adhesion. This effect is due to adsorption of cell adhesion proteins from body fluids or cell culture media onto the polymeric materials, with cells bound to the adsorbed proteins rather than directly interacting with the polymeric biomaterial. Polymeric materials are all somewhat hydrophobic (relative to water, all materials are hydrophobic), while proteins, present in all body fluids, are generally amphiphillic. Thus, proteins will adsorb to polymers, exposing hydrophobic regions to the polymeric material and hydrophilic regions to the surrounding fluid to create a more thermodynamically favorable state. Electrostatic interactions can also be important because proteins have both negatively and positively charged functionality.

Because protein adsorption is driven by hydrophobic and electrostatic interactions, hydrophilic, uncharged materials usually support the least protein adsorption and thus cell adhesion. Nonionic hydrogels, cross-linked networks of nonionic, water-soluble polymers, are relatively cell nonadhesive due to their very hydrophilic and electroneutral nature. Hydrogels generally have rather poor mechanical properties, however, so a great deal of work has focused on synthesizing polymeric materials with surface characteristics similar to a hydrogel but different bulk properties, either by grafting a hydrophilic material to the surface (Gombotz et al, 1989) or by using polymer matrices with both hydrophobic and hydrophilic regions such that hydrophilic moieties are present at the surface when the material is in contact with water (Drumheller and Hubbell, 1994a; Silver et al, 1994).

Immobilization of Proteins onto Polymers for Biospecific Cell Adhesion

Proteins may be immobilized upon polymeric material surfaces to confer biological recognition. Adhesion proteins as well as antibodies have been employed. While immobilized proteins may be used to great advantage to impart biological properties to a synthetic polymer, the nature of the protein may present a number of difficulties. Protein structures are rather delicate and require mild processing conditions without exposure to organic solvents; this is often difficult to achieve when working with synthetic polymers. Furthermore, proteins generally are not stable following incorporation into a polymeric material due to denaturation and proteolytic degradation. If the protein is adsorbed to the polymeric surface rather than covalently immobilized, desorption of the protein will also render the bioactive nature unstable.

Covalent immobilization of cell adhesion proteins has been utilized to render cell nonadhesive polymer materials cell adhesive. Collagen and fibronectin have been covalently attached to poly(vinyl alcohol) hydrogels; 12 μg/cm^2 of collagen was achieved using cyanogen bromide activation (Kobayashi and Ikada, 1991a). Corneal epithelial cells were able to adhere and proliferate when seeded onto poly(vinyl alcohol) hydrogel films with fibronectin immobilized on the surface (Kobayashi and Ikada, 1991b); the films were relatively nonadhesive without protein immobilization. Moreover, pretreatment of expanded polytetrafluoroethylene vascular grafts with fibronectin enhanced endothelialization in vivo (Seeger and Klingman, 1988).

Antibodies have also been utilized as a means to achieve cell type-specific biomaterials. Adhesion of endothelial cells onto polyethylene with adsorption of $F(ab')_2$ fragments of monoclonal antibodies directed against endothelial cell-specific membrane antigens was similar to that observed on adsorbed fibronectin but without increased reactivity towards platelets (Dekker et al, 1991); however, endothelial cells grown upon these antibody-coated surfaces proliferate only in the presence of endothelial cell growth factor (Poot et al, 1991).

Incorporation of Oligopeptides to Facilitate Biospecific Cell Adhesion

A number of short peptide sequences derived from adhesion proteins have been identified that are able to bind to cell-surface receptors and mediate cell adhesion with affinity similar to that obtained with intact proteins. Oligopeptides may be preferable to whole proteins, as they are not subject to denaturation and may be less prone to proteolysis. Some of these, along

Table 1. Cell binding domain sequences of extracellular matrix proteins

Sequence	Protein
RGD	Fibronectin, Vitronectin, Laminin, Collagen, Fibrinogen, Thrombospondin, von Willebrand's factor, entactin, tenascin
YIGSR	Laminin
IKVAV	Laminin
LRE	Laminin
LGTIPG	Laminin
PDSGR	Laminin
RNIAEIIKDA	Laminin
REDV	Fibronectin
DGEA	Collagen
VTXG	Thrombospondin
VGVAPG	Elastin

with their biological sources, are shown in Table 1. A number of methods for incorporating oligopeptides into materials have been investigated, including physicochemical adsorption, chemisorption, and covalent attachment. In this chapter, only covalent attachment will be addressed. The nature of the synthetic polymer substrate to which a peptide is attached appears to be important in a cell's ability to interact with the peptide. For instance, the hydrophobicity and the steric hindrance at the surface appear to be important; spacer arms may be required on some materials. In some cases, the orientation of the peptide may be important as well. Oligopeptides have been covalently attached to both two-dimensional substrates and three-dimensional polymer matrices.

RGD Peptides

The sequence Arg-Gly-Asp (RGD) is found in many cell adhesion proteins and binds to integrin receptors on a wide variety of cell types. RGD interacts with a number of integrin receptors and thus is capable of binding to most cell types (Humphries, 1990). Moreover, the short oligopeptides based on RGD are able to induce cell adhesion with nearly the same affinity as the intact cell adhesion proteins. This was the first integrin binding site to be identified and has been most extensively studied for tissue engineering applications.

A number of studies have been performed with RGD-containing peptides grafted to polymers. The oligopeptide Gly-Arg-Gly-Asp-Tyr (GRGDY) was immobilized onto a polymer-modified glass substrate, and the tyrosine residue was radioiodinated to permit correlation of cell adhesion and function with peptide density (Massia and Hubbell, 1990; 1991). A peptide density of

10 fmol/cm^2 was sufficient to support adhesion and spreading of fibroblast cells, clustering of integrin receptors, and organization of actin stress fibers; at 1 fmol/cm^2 cells were fully spread, but the morphology of the stress fibers was abnormal, and the cells did not form focal contacts. In addition, Gly-Arg-Gly-Asp-Ser (GRGDS) has been grafted onto semi-interpenetrating polymer networks containing polyethylene glycol (PEG), a very hydrophilic polymer (Drumheller and Hubbell, 1994b; Drumheller et al, 1993). In this system, the PEG chains dominated the surface, thus creating a material with good mechanical properties but with extremely low protein adsorption and cell adhesion. When the GRGDS peptide was grafted onto these materials, adhesion of fibroblasts was equivalent to that observed upon tissue culture polystyrene, while, when several control peptides (similar to RGD, but with single amino acid changes) were utilized, no cell adhesion occurred in the absence of serum proteins.

In addition to grafting peptides onto the surface of a polymeric material, peptides can be incorporated into polymer matrices to allow cell adhesion throughout the material, e.g., to allow the possibility of polymer erosion while still permitting the display of a bioadhesive surface. A biodegradable copolymer of polylysine and polylactic acid was used for the attachment of RGD peptides (Barrera et al, 1993). These types of materials can be processed into highly porous sponges to allow cells to permeate into the matrix and adhere throughout (Mikos et al, 1994; Mooney et al, 1995). RGD peptides have also been incorporated into hydrogels formed by polymerizing 8 kDa PEG diacrylate, cell nonadhesive base materials, and high degrees of cell attachment and spreading have been achieved. With PEG hydrogels, a 3.4 kDa PEG spacer arm was required to attain cell adhesion, presumably due to steric hindrance imposed by the highly mobile PEG chains (West et al, 1995).

Bioactive polymers containing RGD sequences have been used in several in vivo applications. RGD peptides coupled to N-(2-hydroxypropyl) methacrylamide hydrogels encouraged growth of glial tissue into the hydrogels when implanted in the striata of rat brains (Woerly et al, 1995). It may be possible to use such hydrogels to support and guide axonal regeneration in severed peripheral nerves. Further, clinical trials have been undertaken to investigate the efficacy of hydrogels containing RGD peptide sequences for promoting healing of diabetic ulcers (Steed et al, 1995). In this study, the percentage of patients whose ulcers healed completely was increased fourfold compared to the placebo group, and the rate of healing was significantly increased compared to the placebo group as well.

Laminin-Derived Peptides

Laminin is a major component of basement membranes and binds to both integrin and nonintegrin receptors. Laminin contains the ubiquitous RGD sequence but also several other unique cell-binding peptides. Tyr-Ile-Gly-Ser-Arg (YIGSR) is a peptide derived from laminin that has been shown to promote cell adhesion (Graf et al, 1987). YIGSR binds to a 67 kDa nonintegrin laminin receptor found in many cell types (Mecham, 1991). Quantitative studies have been performed with YIGSR-containing peptides both adsorbed and covalently immobilized onto a polymer-modified glass substrate that does not support vigorous cell adhesion (Massia et al, 1993). Cell spreading occurred only when the oligopeptide was covalently immobilized, suggesting that conformational constraint of the ligand is necessary for optimal interactions. Similarly, cyclic YIGSR peptides are biologically more active than their linear counterparts (Kleinman, 1989). Moreover, the orientation of the YIGSR peptide appears to influence cell adhesion. Endothelial cells were cultured on substrates with Gly-Tyr-Ile-Gly-Ser-Arg-Gly (GY-IGSRG) that was immobilized either through its N-terminus or its C-terminus (Hubbell et al, 1992). Both cases supported cell adhesion and spreading, but only cells cultured upon the amine-linked peptide were able to withstand a shear force of 20 dyne/cm^2.

Laminin is a potent stimulator of neurite outgrowth, and laminin-derived peptides may have tissue engineering applications in the development of polymeric materials to support and guide peripheral nerve regeneration. The laminin-derived peptide Ile-Lys-Val-Ala-Val (IKVAV) promotes adhesion of neuronal cells via nonintegrin receptors and stimulates collagenase activity (Kanemoto et al, 1990; Sephel et al, 1989), which may be important in wound healing as migrating cells must degrade matrix proteins (e.g., collagen) in their pathway. Laminin may represent a future source of cell-type specific adhesion ligands, as nonintegrin receptors are more likely to be cell selective. For example, the sequence Leu-Arg-Glu (LRE) from laminin apparently supports adhesion of ciliary motorneurons but not of other neuronal cell types or fibroblasts (Hunter et al, 1991).

REDV Peptides

The peptide sequence REDV, derived from fibronectin, has been reported to support adhesion of endothelial cells while not allowing adhesion of other vascular or vessel wall-derived cells, such as platelets, fibroblasts, or smooth muscle cells (Hubbell et al, 1991). GREDVY was immobilized onto polyethylene terepthalate that had been surface modified with polyethylene glycol such that it was cell nonadhesive. The peptide was present at a concentration less than 20 fmol/cm^2, yet endothelial cells exhibited complete spreading, indicating the potency of this ligand. This sequence may have an important medical application in the creation of vascular grafts that support endothelial cell adhesion, while limiting adhesion of blood platelets, smooth muscle cells, and fibroblasts, all of which can cause failure of a vascular graft. The

GREDVY sequence has been utilized to achieve selective adhesion of endothelial cells onto vascular grafts (Holt et al, 1994). The peptide was attached to Dacron vascular prosthetic material that was impregnated with polyethylene oxide to render it cell nonadhesive. Attachment of GREDVY increased endothelial cell adhesion while decreasing fibroblast adhesion in vitro. This represents the first report of the use of peptide sequences in bioactive polymers to support adhesion of a single cell type.

Proteoglycan-binding Peptides

Proteoglycans, a basement membrane component, are proteins that are grafted with polysaccharide-based glycosaminoglycan side chains, and proteoglycans serve as low affinity adhesion receptors. Because most glycosaminoglycans are sulfated, these blocks on the proteoglycans bear a net negative charge, and the binding of glycosaminoglycans to cell adhesion proteins appears to be based on ionic interactions, as highly sulfated species bind more tightly (Ruoslahti, 1988). Glycosaminoglycans interact with regions of clustered positive charges, such as the prototypical sequence XBBXBX, where X is a hydrophobic amino acid and B is a positively charged basic amino acid, namely arginine or lysine (Cardin and Weintraub, 1989). Proteoglycan-binding sequences of this genre have been identified on fibronectin, vitronectin, and laminin. Proteoglycan-binding sequences may be important in addition to more classical receptor-binding sequences on materials for tissue engineering, as they are thought to act in concert with binding by integrins and may be necessary for the maintenance of normal phenotypes (Saunders and Bernfield, 1988).

INCORPORATION OF CARBOHYDRATE MOIETIES FOR BIOSPECIFIC CELL ADHESION

Binding to the Asialoglycoprotein Receptor

Monosaccharide structures have been incorporated into polymers to promote biospecific cell adhesion. Because damaged liver tissue will regenerate itself, there has been a drive to develop bioactive polymer materials that are selectively adherent for hepatocytes. The asialoglycoprotein receptor is found on hepatocytes and normally functions to clear certain protein complexes from the circulation, but can be engineered for use as a cell adhesion ligand (Geffen and Speiss, 1992). N-acetylglucosamine has been immobilized onto polystyrene to achieve selective adhesion of avian hepatocytes

(Gutsche et al, 1994), and lactose has been immobilized onto polystyrene for selective adhesion of mammalian hepatocytes (Kobayashi et al, 1992; 1994).

Lectin Domains

Lectins are carbohydrate-binding cell adhesion proteins with sugar-specific binding properties. Lectins may be able to serve as ligands for cell-type selective adhesion onto bioactive polymers. The lectin Ulex europaeus I (UEA I) has a high affinity for endothelial cell surface glycocoproteins. When UEA I was covalently immobilized onto polyethylene terephthalate, endothelial cell attachment was increased 100-fold, whereas adhesion of monocytes, smooth muscle cells, and fibroblasts was reduced (Ozaki et al, 1993). These results imply that carbohydrate interactions may be useful for achieving cell-specific adhesion, and UEA I may be an effective ligand to promote endothelialization of vascular grafts in vivo.

Selectins are lectin-like cell-surface receptors that bind to specific oligosaccharide structures, namely sialylated fucosylated lactosamine oligosaccharides bound to proteins such as the sialyl Lewis X and sialyl Lewis A structures (Varki, 1994). Liposomes containing sialyl Lewis X glycolipid have been shown to bind specifically to embryonal carcinoma cells in vitro (Eggens et al, 1989). Selectin ligands have not been employed in the development of bioactive polymers for tissue engineering as of yet, but may represent an important means to achieve biospecific cell adhesion.

PHARMACOLOGICAL MODIFICATION OF POLYMERIC BIOMATERIALS

Bioactive materials for tissue engineering applications may be modified with species that alter cell function as well as cell adhesion ligands. Such materials may be useful to increase cell proliferation or alter enzyme cascades such as the complement cascade or the coagulation cascade. The species incorporated may include proteins, peptides, or charged groups, to name a few. Two applications are discussed below: immobilization of heparin to locally alter the coagulation cascade and binding of growth factors to increase cell proliferation and maintain cell phenotypes.

Modification with Heparin

Cardiovascular biomaterial surfaces have been modified with heparin to obtain localized anticoagulation. Heparin is a natural polysaccharide that catalyzes the inactivation of thrombin by antithrombin III. Heparin has been immobilized to poly(vinyl alcohol), and the patency of ex vivo shunts coated with this polymer was evaluated in a canine model (Ip and Sefton, 1989). At low flow rates, heparinized tubes remained patent for greater than 72 min, while control tubes became occluded after approximately 30 min. End grafting of heparin onto polyurethane tubes increased the mean duration of patency from 3.5 days to 25.9 days in a canine arteriovenous shunt model (Arnander, 1987).

Because heparin must complex with antithrombin III in a specific structure, the conformation of heparin after incorporation into a bioactive polymer is crucial in determining its efficacy. Chemical schemes that aim to reduce steric hindrance of the heparin chain appear to provide the greatest efficacy: these have included end-grafting (Grainger et al, 1988; 1990) and the use of 3400 Da polyethylene glycol spacer arms (Byun et al, 1994).

Immobilization of Growth Factors

Polypeptide growth factors may be incorporated into bioactive polymers to increase cell proliferation, migration, or differentiation. Growth factors are generally small proteins (less than 200 amino acids) and bind to cell-surface receptors to stimulate cellular activity via tyrosine kinase phosphorylation. Growth factor classes include fibroblast growth factors (FGFs), insulin-like growth factors (IGFs), epidermal growth factors (EGFs), and platelet-derived growth factors (PDGFs).

Covalent immobilization of insulin onto polyurethane membranes significantly stimulated endothelial cell proliferation relative to native polyurethane or polyurethane with immobilized collagen. Moreover, coimmobilization of both collagen and insulin further accelerated proliferation, and endothelial cells grown on this substrate maintained their normal morphology and the ability to secrete prostacyclin for 9 months (Liu et al, 1993). Insulin has also been covalently immobilized onto poly(methyl methacrylate). Immobilized insulin stimulated growth of fibroblasts, and coimmobilization with fibronectin further increased cell proliferation (Ito et al, 1993). Expanded polytetrafluoroethylene (ePTFE) vascular grafts pretreated with releasable fibroblast growth factor type 1 supported enhanced endothelialization and capillary ingrowth in a rabbit model (Greisler et al, 1992). Every FGF-treated graft in this study displayed a confluent endothelium, while neither the native ePTFE nor those modified with fibrin glue did, indicating that the FGF-treatment did in fact increase the rate of graft endothelialization in vivo. Endothelialization of vascular grafts may reduce graft failure caused by thrombosis and neointima formation.

Utilization of growth factors to impart bioactive properties to polymeric materials is plagued by the same inherent problems as immobilization of whole cell adhesion proteins, namely potential instability due to denaturation and proteolytic degradation. While they have not been utilized thus far in the development of bioactive materials, several short peptide sequences have been identified that can specifically bind to growth factor receptors and elicit appropriate biological responses, similar to the various peptides used to mimic cell adhesion proteins. Linear and cyclic synthetic peptides that corresponded to the receptor binding site of EGF were able to bind to the receptor, albeit with significantly reduced affinity relative to the native protein, and to stimulate proliferation of human fibroblasts (Komoriya et al, 1984). The synthetic peptide corresponding to basic FGF residues 106–115 was shown to bind to both the FGF receptor and to heparin and further to stimulate DNA synthesis in fibroblasts (Baird et al, 1988).

CONCLUSIONS

The use of bioactive polymers for cell growth scaffolds may enhance tissue ingrowth and cell proliferation. Depending on the requirements imposed by the clinical situation, materials may be designed to display broad affinity for cells or to promote the adhesion and phenotypic expression of selected cell types. This has implications for the development of superior medical devices and engineered tissues for transplant, e.g., vascular graft materials and materials for liver regeneration.

References

Arnander C (1989): Enhanced patency of small-diameter tubings after surface immobilization of heparin fragments. *J Biomed Mater Res* 23:285–94

Baird A, Schubert D, Ling N, Guillemin R (1988): Receptor-binding and heparin-binding domains of basic fibroblast growth factor. *Proc Natl Acad Sci USA* 85:2324–8

Barrera DA, Zylestra E, Lansbury PT, Langer R (1993): Synthesis and RGD peptide modification of a new biodegradable polymer- poly(lactic acid-co-lysine). *J Am Chem Soc* 115:11010–1

Byun Y, Jacobs HA, Kim SW (1994): Heparin surface immobilization through hydrophilic spacers: thrombin and antithrombin III binding kinetics. *J Biomater Sci Polym Edn* 6:1–13

Cardin AD, Weintraub HJR (1989): Molecular modeling of protein-glycosaminoglycan interactions. *Arteriosclerosis* 9:21–32

Cima LG, Ingber DE, Vacanti JP, Langer R (1990): Hepatocyte culture on biodegradable polymeric substrates. *Biotech Bioeng* 38:145–58

Dekker A, Poot AA, van Mourik JA, Workel MPA, Beugling T, Bantjes A, Feijen J, van Aken WG (1991): Improved adhesion and proliferation of human endothelial cells on polyethylene precoated with monoclonal antibodies against cell membrane antigens and extracellular matrix proteins. *Thromb Haem* 66:715–24

Drumheller PD, Hubbell JA (1994a): Phase-mixed poly(ethylene glycol) poly (trimethylolpropane triacrylate) semiinterpenetrating polymer networks obtained by rapid network formation. *J Polym Sci Polym Chem* 32:2715–25

Drumheller PD, Hubbell JA (1994b): Polymer networks with grafted cell adhesion peptides for highly biospecific cell adhesive substrates. *Analyt Biochem* 222:380–8

Drumheller PD, Elbert DE, Hubbell JA (1993): Multifunctional poly(ethylene glycol) semi-interpenetrating polymer networks as highly selective adhesive substrates for bioadhesive peptide grafting. *Biotech Bioeng* 43:772–80

Eggens I, Fenderson B, Toyokuni T, Dean B, Stroud M, Hakamori S (1989): Specific interactions between Lex and Lex determinants. *J Biol Chem* 264:9476–84

Freed LE, Grande DA, Lingbin Z, Emmanuel J, Marquis JC, Langer R (1994): Joint resurfacing using allograft chondrocytes and synthetic biodegradable polymer scaffolds. *J Biomed Mater Res* 28:891–9

Geffen I, Speiss M (1992): Asialoglycoprotein receptor. *Int Rev Cytol* 137B:181–219

Gombotz WR, Guanghui W, Hoffman AS (1989): Immobilization of poly(ethylene oxide) on poly(ethylene terepthalate) using a plasma polymerization process. *J Appl Poly Sci* 37:91–107

Graf J, Ogle RC, Robey FA, Sasaki M, Martin GR, Yamada Y, Kleinman HK (1987): A pentapeptide from the laminin B1 chain that mediates cell adhesion and binds the 67000 laminin Da receptor. *Biochemistry* 26:6896–900

Grainger D, Kim SW, Feijen J (1988): Poly(dimethylsiloxane)-poly(ethylene oxide)-heparin block copolymers: Synthesis and characterization. *J Biomed Mater Res* 22:231–49

Grainger DW, Knutson K, Kim SW, Feijen J (1990): Poly(dimethylsiloxane)-poly(ethylene oxide)-heparin block copolymers: Surface characterization and in vitro assessments. *J Biomed Mater Res* 24:403–31

Greisler HP, Cziperle DJ, Kim DU, Garfield JD, Petsikas D, Murchan PM, Applegren EO, Drohan W, Burgess WH (1992): Enhanced endothelialization of expanded polytetrafluoroethylene grafts by fibroblast growth factor type 1 pretreatment. *Surgery* 112:244–54

Gutsche AT, Parsons-Wigerter P, Chand D, Saltman WM, Leong KW (1994): N-acetylglucosamine and adenosine derivatized surfaces for cell culture: 3T3 fibroblast and chicken hepatocyte response. *Biotech Bioeng* 43:801–9

Holt DB, Eberhart RC, Prager MD (1994): Endothelial cell binding to Dacron modified with polyethylene oxide and peptide. *ASAIO J* 40:858–63

Hubbell JA, Massia SP, Desai NP, Drumheller PD (1991): Endothelial cell-selective tissue engineering in the vascular graft via a new receptor. *Bio/Technology* 9:568–72

Hubbell JA, Massia SP, Drumheller PD, Herbert CB, Lyckman AW (1992): Bioactive and cell-type selective polymers obtained by peptide grafting. *Polym Mater Sci Eng* 203:16

Humphries MJ (1990): The molecular basis and specificity of integrin-ligand interactions. *J Cell Sci* 97:585–92

Hunter DD, Cashman N, Morrisvalero R, Bulock JW, Adams SP, Sanes JR (1991): An LRE (leucine arginine glutamate)-dependent mechanism for adhesion of neurons to S-laminin. *J Neurosci* 11:3960–71

Ip WP, Sefton MV (1989): Patency of heparin-PVA coated tubes at low flow rates. *Biomaterials* 10:313–17

Ito Y, Inoue M, Liu SQ, Imanishi Y (1993): Cell growth on immobilized cell growth factor: enhancement of fibroblast growth by immobilized insulin and/or fibronectin. *J Biomed Mater Res* 27:901–7

Kanemoto T, Reich R, Royce L, Greatorex D, Adler SH, Shiraishi N, Martin GR, Yamada Y, Kleinman HK (1990): Identification of an amino-acid sequence from the laminin-A chain that stimulates metastasis and collagen IV production. *Proc Natl Acad Sci USA* 87:2279–83

Kleinman HK, Graf J, Iwamoto Y, Sasaki M, Schasteen CS, Yamada Y, Martin GR, Robey FA (1989): Identification of a 2nd Active-Site in Laminin for Promotion of Cell-Adhesion and Migration and Inhibition of In vivo Melanoma Lung Colonization. *Arch Biochem Biophy* 272:39–45

Kobayashi H, Ikada Y (1991): Covalent immobilization of proteins onto the surface of poly(vinyl alcohol) hydrogel. *Biomaterials* 12:747–51

Kobayashi H, Ikada K (1991): Corneal cell adhesion and proliferation on hydrogel sheets bound with cell-adhesive proteins. *Curr Eye Res* 10:899–908

Kobayashi A, Kobayashi K, Akaike T (1992): Control of adhesion and detachment of parenchymal liver cells using lactose-carrying polystyrene as substratum. *J Biomat Sci Polym Ed* 3:499–508

Kobayashi K, Kobayashi A, Akaike T (1994): Culturing hepatocytes on lactose-carrying polystyrene layer via asialoglycoprotein receptor-mediated interactions. *Meth Enzymol* 247:409–418

Komoriya A, Hortsch M, Meyers C, Smith M, Kanety H, Schlessinger J (1984): Biologically active synthetic fragments of epidermal growth factor: Localization of a major receptor-binding region. *Proc Natl Acad Sci USA* 81:1351–5

Liu SQ, Ito Y, Imanishi Y (1993): Cell growth on immobilized growth factor: Covalent immobilization of insulin, transferrin, and collagen to enhance growth of bovine endothelial cells. *J Biomed Mater Res* 27:909–15

Massia SP, Hubbell JA (1990): Covalent surface immobilization of Arg-Gly-Asp and Tyr-Ile-Gly-Ser-Arg- containing peptides to obtain well-defined cell-adhesive substrates. *Analyt Biochem* 187:292–301

Massia SP, Hubbell JA (1991): An RGD spacing of 440 nm is sufficient for integrin $\alpha_v\beta_3$-mediated fibroblast spreading and 140 nm for focal contact and stress fiber formation. *J Cell Biol* 114:1089–1100

Massia SP, Rao SS, Hubbell JA (1993): Covalently immobilized laminin peptide Tyr-Ile-Gly-Ser-Arg (YIGSR) supports cell spreading and co-localization of the 67-kilodalton laminin receptor with α-actinin and vinculin. *J Biol Chem* 268:8053–9

Mecham RP (1991): Laminin receptors. *Ann Rev Cell Biol* 7:71–91

Mikos AG, Thorsen AJ, Czerwonka LA, Bao Y, Langer R, Winslow DN, Vacant JP (1994): Preparation and characterization of poly(L-lactic acid) foams. *Polymer* 35:1068–77

Mooney DJ, Park S, Kaufmann PM, Sano K, McNamara K, Vacanti JP, Langer R (1995): Biodegradable sponges for hepatocyte transplantation. *J Biomed Mater Res* 29:959–65

Ozaki CK, Phaneuf MD, Hong SL, Quist WC, LoGerfo FW (1993): Glycoconjugate mediated endothelial cell adhesion to Dacron polyester film. *J Vasc Surg* 18:486–94

Pathak CP, Sawhney AS, Hubbell JA (1992): Rapid photopolymerization of immunoprotective gels in contact with cells and tissues. *J Am Chem Soc* 114:8311–2

Poot AA, Spijkers J, van Mourik JA, Hoentjen GJ, Engbers GHM, Beugeling T,

Bantjes A, Feijen J, van Aken WG (1991): Endothelialization of vascular graft materials by surface coating with a monoclonal antibody against the endothelial membrane protein endoglin. *Thromb Haem* 69:572

Ruoslahti E (1988): Structure and biology of proteoglycans. *Ann Rev Cell Biol* 4:229–55

Saunders S, Bernfield M (1988): Cell-surface proteoglycan binds mouse mammary epithelial-cells to fibronectin and behaves as a receptor for interstitial matrix. *J Cell Biol* 106:423–30

Seeger JM, Klingman N (1988): Improved in vivo endothelialization of prosthetic grafts by surface modification with fibronectin. *J Vasc Surg* 8:476–82

Sefton MV, Broughton RL, Sugamori ME, Mallabone CL (1987): Hydrophilic polyacrylates for the microencapsulation of fibroblasts or pancreatic islets. *J Cont Rel* 6:177–87

Sephel GC, Tashiro KI, Sasaki M, Greatorex D, Martin GR, Yamada Y, Kleinman HK (1989): Laminin-A chain synthetic peptide which supports neurite outgrowth. *Biochem Biophys Res Comm* 162:821–9

Silver JH, Myers CW, Lim F, Cooper SL (1994): Effect of polyol molecular-weight on the physical-properties and hemocompatibility of polyurethanes containing poly (ethylene oxide) macroglycols. *Biomaterials* 15:695–704

Steed DL, Ricotta JJ, Prendergast JJ, Kaplan RJ, Webster MW, McGill JB, Schwartz SL (1995): Promotion and acceleration of diabetic ulcer healing by arginine-glycine-aspartic acid (RGD) peptide matrix. *Diabetes Care* 18:39–46

Varki A (1994): Selectin ligands. *Proc Natl Acad Sci USA* 91:7390–7

West JL, Hern DL, Hubbell JA (1995): Polymers in medicine: manipulating wound healing. *Poly Mat Sci Eng* 210:222

Woerly S, Laroche G, Marchand R, Pato J, Subr V, Ulbrich K (1995): Intracerebral implantation of hydrogel-coupled adhesion peptides: Tissue reaction. *J Neural Transplant Plast* 5:245–55

6

DRUG DELIVERY RELATED TO TISSUE ENGINEERING

KAM W. LEONG

INTRODUCTION

Success of tissue engineering depends on many factors. Host response to the scaffold, template, or the cell-seeded implant is one of the critical parameters. The concept of using pharmacological means to redirect or effect a favorable biological response during the course of tissue reconstruction is appealing. Examples abound on the incorporation of a drug delivery function into prostheses (Greco, 1994). For instance, antimicrobial agents have been embedded in bone cement or catheters to combat infection, and antithrombogenic agents have been incorporated into or conjugated to the surface of vascular grafts to minimize thrombosis. Recent discoveries of potent polypeptides that can influence the course of cell maturation and development accentuate the importance of considering delivery of these factors in designing a scaffold for tissue regeneration. A prime example is bone regeneration.

One of the critical elements for tissue engineering of bone is the provision of a regulatory signal for bone initiation (Reddi, 1994). Bone morphogenetic proteins (BMPs), members of the TGF-b superfamily, can provide the signals. BMP-3, for instance, stimulates alkaline phosphatase activity, collagen synthesis of osteoblasts, and proteoglycan synthesis of chondrocytes. Beyond the initiation stage, other growth factors such as TGF-β1 and TGF-β2, in addition to BMPs, may be needed to promote and maintain the osteogenic phenotypes. Optimal bone repair, therefore, may need the timely delivery of a combination of these growth and differentiation factors in

Synthetic Biodegradable Polymer Scaffolds
Anthony Atala and David Mooney, Editors
Robert Langer & Joseph P. Vacanti, Associate Editors
© 1997 Birkhäuser Boston

addition to the choice of an appropriate substratum to mimic the extracellular matrix.

Ever since the finding that central axons can be induced to regenerate along a graft, attempts have been made to repair peripheral nerve injury by bridging the gap at a lesion site through a tubular guide (Brunelli et al, 1994). However, success has been limited in that only a small proportion of injured axons can be regenerated. Recent findings suggest that regeneration of neurons is inhibited by multiple signals (Nash and Pini, 1996). Myelin-associated glycoprotein has been shown to promote outgrowth of neonatal dorsal root ganglion cells but inhibit that of adult cells and cerebellar neurons. Another protein isolated from adult gray matter has also been shown to inhibit the growth of central and peripheral neurons *in vitro*. A potential strategy to promote nerve regeneration is hence to locally deliver neutralizing antibodies to these inhibitors by the graft. Alternatively, growth factors can be delivered locally to promote axonal regeneration. For instance, exogenous brain-derived neurotrophic factor and neurotrophin-3 have been shown to enhance propriospinal axonal regeneration in a rat model (Xu et al, 1995).

When the drug of interest is a protein, delivery through cells is possible. Small diameter vascular grafts (<5 mm) have always been plagued by thrombotic occlusion. Seeding of the graft surface with endothelial cells, possibly transduced with foreign genes, has been proposed to improve the patency of the grafts following arterial reconstruction (Williams and Jarrell, 1996). A similar utility of such cell-seeded conduit is proposed in an artificial kidney design (Cieslinski and Humes, 1994). The glomerular filtration function is currently simulated by a dialysis unit typically made of polysulphone hollow fibers. However, protein deposition on the semipermeable membrane and thrombotic occlusion of fibers are frequent problems. Local sustained release of anticoagulant factors, such as heparin, has been proposed to minimize thrombogenesis on the surface. The antithrombogenic effect generally, however, is transient because the supply of the anticoagulant is limited. It has been proposed that the filtration surfaces be seeded with endothelial cells, possibly transduced with anticoagulation factors, such as plasminogen activator, to minimize thrombosis. In either case, success of the artificial tissue is aided by judicious integration of drug delivery.

This brief review will first discuss the basic mechanisms of drug release from a polymeric drug-carrier and then highlight the cellular approaches of drug delivery.

POLYMERIC CONTROLLED DRUG DELIVERY

Controlled release technology can be broadly categorized as liposomal, electromechanical, and polymeric. In relation to tissue engineering, the polymeric technology is the most relevant. The types of polymer used for controlled release can be biodegradable, nonbiodegradable, and soluble. Conceivably all three types are applicable in tissue engineering. The first two types serve as a solid carrier, existing in the form of microspheres, matrices, and membranes to deliver the bioactive agents. Soluble polymers can be used as a carrier covalently linked to the drug, and the resulting complex is localized in the implant for delivery. The design of controlled delivery systems has called on many ingenious strategies, involving elaborate synthesis of the polymeric carriers and the drug-polymer conjugates. Extra constraints imposed on the scaffold configuration, for example, a highly porous architecture to allow cell ingrowth, may also require intricate fabrication design.

The drug-release mechanism can be physical, chemical, or enzymatic in nature. For nonbiodegradable matrix and membrane devices, release is by diffusion through the polymer phase and is driven by the concentration gradient. The release can also be triggered by osmotic pressure or matrix swelling. For matrices that are biodegradable or containing drug conjugates, release is controlled by the hydrolytic or enzymatic cleavage of the relevant chemical bonds. Even in such biodegradable systems, diffusion of the reactant, which can be water or the enzyme, and the liberated drug molecules may still be rate-limiting. Only a qualitative discussion is presented here for understanding the release patterns ensuing from the more common delivery systems. A quantitative understanding of the controlled release behavior is extensively covered in an excellent book (Fan, 1989).

Diffusion-controlled release mechanism

Under this category, the release may be controlled by drug diffusion or solvent penetration. The diffusion-controlled systems can be distinguished into reservoir and matrix devices. In the former case, the drug reservoir is encapsulated by a polymeric membrane. The drug core can be in the solid or the liquid state, and the membrane can be microporous or nonporous. If the drug core is maintained at a saturated state, the transport of drug molecules across the membrane will be steady, as the driving force for permeation is constant. This constant release rate is termed zero-order release, to denote the independence of the release rate on drug concentration (i.e., $dC/dt = kC^n$, where n is 0). Such a zero-order release kinetics requires the drug core to remain in a solid or suspension state. The saturate state would be difficult to maintain if the drug has high water solubility. Even if all the requirements for constant release are met, the release is generally not constant in the initial and end periods. Depending on the history of the device, a lag time or a burst effect may be observed. If the membrane encapsulating the drug core is devoid of drug molecules at the time of release, an induction period is needed to saturate the membrane.

When the device is placed in a releasing medium immediately after fabrication, it takes a certain time for the system to reach a steady state. If, on the other hand, the drug molecules have been accumulating in the membrane as a result of storage, the initial release rate will be higher than the steady-state value. The time it takes for the release rate to reach the steady state varies with the nature and the thickness of the membrane. Towards the end stages of release, the rate will decline as the drug concentration in the core drops below the saturation limit.

The simplest and most widely used controlled release devices are matrix systems in which the drug is dissolved or dispersed in the polymer. One characteristic of these system is a release rate decreasing with time as a result of diminishing concentration gradient across the device, which, simplistically, can be viewed as a consequence of increasing diffusion distance for the drug solutes to travel from the core to the surface. This release kinetics is often described as first order (i.e., $dC/dt = kC^n$, where n is 1). The cumulative amount of drug released is proportional to the square root of elapsed time. It is possible to obtain a more steady release by using multi-lamination to create a nonuniform drug distribution across the device. For instance, a drug concentration that increases towards the center of the device can in principle maintain the concentration gradient constant as release progresses. The complicated fabrication procedure, however, is not likely to make this approach widely practical. This design is also valid only if there is no equilibration of drug concentration in the matrix before use. In addition to permeation through the polymer phase, the drug molecules can also diffuse through channels created by dissolution of the drug phase. In fact, this is the only avenue by which macromolecules and drugs that have low permeability can be released. Since the extent and size of the pores and channels created are determined by the drug incorporated, the drug loading level and particle size of the drug solutes have a profound influence on the release kinetics. Below a critical loading level, there may not be enough interconnected channels, and some of the isolated drug particles may be trapped inside the matrix. When dealing with precious and potent growth factors in tissue engineering applications, the protein loading level may be low. To effect release, inert filler molecules such as sucrose, inulin, or albumin can be mixed with the growth factor to enhance the interconnectivity of the channels.

The release behavior of these diffusion-controlled systems is highly dependent on the physical properties of the drug. In addition to loading level and particle size of the solutes, the drug solubility in the polymer and drug diffusivity in the polymer phase are both important parameters. The shape of the device, which determines the surface area available and the path length for diffusion, is also important.

Poly(dimethysiloxane) and poly(ethylene-vinyl acetate) (EVAc) are two prominent examples of nonbiodegradable polymers used to construct the reservoir and matrix controlled release devices. The former has been used to deliver contraceptive steroids, whereas EVAc has been used to deliver a large number of bioactive agents, ranging from pilocarpine and progesterone to antibodies and nerve growth factors (Langer, 1990; Langer and Wise, 1986; Leong and Langer, 1987). Nonbiodegradable systems are attractive because their release behavior is predictable, and early termination of treatment is possible provided the removal of the device is not prevented by fibrous tissue reaction. Poly(dimethylsiloxane) and EVAc are attractive drug-carriers because they are inert, invoking minimal tissue response. This property however may be undesirable for tissue engineering if their high hydrophobicity prevents favorable interaction with cells.

Solvent-controlled release mechanism

In solvent-controlled systems, the release of drug is regulated by the permeation of water through the polymer. An osmotic pump is constructed by enclosing a drug core with a semipermeable membrane equipped with an orifice. When placed in water, the device will deliver a volume of drug solution or suspension equal to the water influx. The release rate is then determined by the nature of the membrane, the size of the orifice, and the osmotic activity of the drug core. A constant release rate is obtained if the osmotic pressure across the membrane is kept constant by maintaining a saturated drug core. More recent designs separate the drug and osmotic agent compartments by a flexible membrane. Water imbibed into the osmotic agent compartment flexes the separating membrane to squeeze out drug solution through the orifice in the drug compartment.

Another solvent-activated system depends on the relaxation of the polymer as a result of water absorption. The drug is incorporated into a hydrogel, which in the dry state is glassy. In the presence of water, the hydrogel relaxes into an elastic state which presents little resistance to the diffusion of the drug solutes. The release rate is therefore mainly controlled by the swelling rate of the polymer, which in turn is controlled by the hydrophilicity and cross-linking density of the hydrogel. In protein delivery, ionic interaction and chain entanglement between the protein and the matrix can also retard the release.

The chain mobility and high water content of hydrogels are valuable in bioadhesive designs. Their interactions with mucus lining of the tissue may keep the controlled release device at the target site for a longer period of time. This adhesiveness may be advantageous in wound healing applications. Cross-linked poly(2-hydroxyethyl methacrylate) (PHEMA) and its copolymers with poly(N-vinyl-pyrrolidone), polyacrylamide, and poly(vinyl alcohol) are examples of commonly studied hydrogels (Peppas and Langer, 1994). Natural biopolymers such as collagen, gelatin, hyaluronic acid, and chondroitin sulfate, have also been cross-linked to form hydrogels for drug delivery. These biopolymers are particularly attractive for tissue reconstruction because they are ideal substrates for cell attachment and maintenance of cell functions. They may also be a better carrier than synthetic hydrophobic polymers in preventing the entrapped proteins from denaturing, although systematic studies are needed for substantiation.

The ideal drug release profile is not always a constant delivery. While that is the goal of many delivery systems, a release that is responsive to a stimulus or physiological needs would be superior. Insulin and other hormonal therapies are prime examples. Researchers have relied on autoregulatory and external-stimulatory mechanisms to achieve the goal of responsive delivery. For instance, a glucose-responsive insulin delivery system has been developed based on the complementary and competitive binding of a lectin protein, concanavalin A, with glucose and glycosylated insulin (Jeong et al, 1986). The binding affinity of the glycosylated insulin can be adjusted by varying the sugar moiety and the spacer that links the sugar to the insulin molecule. A different autoregulating approach relies on the change of membrane permeability in response to external glucose concentration (Albin et al, 1986). The system consists of a macroporous membrane such as poly-HEMA enclosing a saturated insulin solution. The membrane contains pendant amine groups and immobilized glucose oxidase. As glucose diffuses into the gel, glucose oxidase catalyzes its conversion to gluconic acid, thereby lowering the pH within the gel microenvironment. The low pH leads to ionization of the pendant amine groups, which in turn increases the swelling of the hydrogel and its permeability to insulin.

Degradation-controlled release mechanism

The two common chemically controlled systems are a biodegradable matrix in which the drug is dispersed and a polymer-drug conjugate with the drug molecules linked to the side chains of the polymer. The dissolution of the matrix is effected by hydrolytic or enzymatic cleavage of the backbone of the polymer. The cleavage can also occur in the cross-linking bonds, rendering soluble an initially cross-linked polymer. Synthetic hydrogels become biodegradable by using hydrolytically labile cross-linkers. Hydrogels composed of natural biopolymers are also biodegradable because their backbone can be cleaved by enzymes even though they are cross-linked by hydrolytically stable bonds. Alternatively, the dissolution can originate from hydrolysis, ionization, or protonation of the side chains of the polymer. If drug release for these chemically controlled systems is governed solely by the biodegradation of the matrix, a constant release rate may be obtained, provided the surface area of the device is maintained constant. However, that represents an idealized case that is rarely realized. In reality the drug molecules can also diffuse through the matrix. This is particularly true if the drug is hydrophilic, which presents a high driving force for diffusion. The release is often intermediate between zero-order and first-order kinetics. A constant release can sometimes be achieved if the diffusion is accompanied by a change in the matrix properties which boosts the diffusivity of the drug. For instance, the diffusivity may increase with time as a result of partial degradation or swelling of the matrix. Such an increase may compensate for the decrease in the concentration gradient and lead to a constant release.

The obvious advantage of a biodegradable system is the elimination of the need to remove the drug-depleted device. In tissue engineering, the concept of a scaffold that can be totally replaced by the host tissue has been widely advocated. The degradation-controlled release systems are thus likely to be the most important in tissue engineering. This system also enjoys the advantage of having release kinetics that are less dependent on the properties of the drugs than are the nonbiodegradable systems. There is no worry that proteins might be trapped in the matrix and less of a hazard that the protein might aggregate, which might happen in a hydrophobic nonbiodegradable matrix.

The release rate of the polymer-drug conjugate, which has been called a pendant system, is dependent on the cleavage of the polymer-bond. If the drug is attached to the polymer via a spacer, the hydrolysis of the polymer spacer and the spacer-drug bonds are both relevant. The spacer approach provides an effective means of controlling the release rate. A spacer may also be necessary in some cases to render the conjugation bond accessible to hydrolysis. Although penetration of water into the matrix and outward diffusion of the cleaved drug molecules constitute part of the rate barrier, the cleavage of the polymer-drug bond as the rate-determining step is preferred for better control. Generally, the release rate drops with time as the drug concentration decreases.

The potential advantage of the pendant system is the high drug loading level. While the drug loading level of biodegradable matrix systems would be generally under 20 weight percent, a higher drug-to-polymer ratio is possible in a pendant system. This may also be the only system that can provide a sustained release, for example, a release period of weeks, for highly hydrophilic drugs. It might be possible to provide a sustained release for highly hydrophilic drugs using other release systems described above if the drug loading is small, for instance, less than one weight percent. At high loading, sustained release will not be possible from a matrix system because the hydrophilic drug creates a high drive force for diffusion as well as swelling of the matrix. Conceptually, the pendant system affords the best chance to retard the release through the use of a relatively stable polymer-drug bond. The disadvantage of the pendant system is that extensive chemical development might be required for each drug, in conjugation, optimization of the release rate, and in ensuring that the integrity of the drug is restored after release.

Implant-Tissue Interface

In theory it is possible to obtain any desirable release pattern by sophisticated designs such as nonuniform drug distribution or lamination. If the drug is only sparingly soluble in water, such as some steroids or Taxol, dissolution of the drug may become one of the rate-determining steps, and the analysis will have to be adjusted accordingly. Moreover, the overall release kinetic consideration is meaningful only if the pharmacokinetic properties of the drug and the *in vivo* environment are taken into account. The qualitative analysis described above applies to a perfect sink scenario in which the released drug is assumed to be diluted infinitely or clear away rapidly. This may not be true *in vivo*, where the turnover of the interstitial fluid may be slow, or a fibrous reaction may take place on the drug-carrier. The increased mass-transfer resistance at the implant-tissue interface will slow down the release.

Intriguing to tissue engineering applications is how would a cell layer affect the release from the scaffold. Attached cells would deposit their own proteins onto the substrate. This might create an extra transport barrier. If the scaffold is biodegradable, the cell layer probably would also slow down the matrix degradation and drug release, unless the matrix is susceptible to enzymatic degradation, in which case enzymes secreted by the cells would accelerate the scaffold decomposition. As tissue organization takes form in the scaffold, the mass transport situation will become increasingly complex. Penetration of water into the scaffold and out-diffusion of the released drug will probably be retarded. Angiogenesis in the generated tissue on the other hand will improve the clearance of the released drug and the scaffold degradation products. Systematic studies remain to be conducted to provide a theoretical framework for analyzing the complex mass transport scenarios in tissue engineering.

Fabrication Considerations

Kinetics of drug release is dependent on the physical state of the polymeric carrier. Different processing conditions can yield matrices that can range from nonporous to macroporous. The effective diffusivity of the drug in the polymer phase will vary accordingly. The crystallinity of the polymeric carrier, which is affected by the fabrication conditions, also plays an important role. Diffusivity in crystalline phase is orders of magnitude lower than that in amorphous phase. The crystalline phase is also resistant to water penetration. For this reason, degradation of polylactide, for example, occurs preferentially in the amorphous phase, and the same polymer might show different degradation rates because of the different fabrication conditions.

The common techniques for preparing drug delivery matrices in the form of films or slabs are solvent casting, compression molding, melt-pressing, and injection molding. Solvent casting is convenient, can be done at low temperature, and requires little equipment. If the drug is also soluble in the solvent for the polymer, it can lead to homogeneous or molecular distribution of the drug in the matrix. Solvent casting is suitable only for polymers with reasonable molecular weights, otherwise the resulting matrix would be fragile. Not all polymers are soluble in suitable organic solvents—suitable in the sense of relatively nontoxic and volatile. This is important because of biocompatibility considerations. There will always be an equilibrium amount of residual solvent in the final matrix. If the polymer is soluble only in high boiling solvents such as dimethylformamide or fluorinated hydrocarbons, then solvent casting is not appropriate. Compression molding and melt pressing are similar in principle but differ in the temperature of processing. Compression molding should be conducted at a temperature above the glass transition point of the polymer. Melt-pressing or transfer molding is conducted at an even higher temperature when the polymer is softened or flowing. Injection molding is to inject a molten mixture of polymer and drug into a die. Injection molding can be a continuous process and is the method of choice for large scale fabrication of drug delivery devices. These molding processes require no contact with organic solvent, but the drug must be thermally stable. There must also be no chemical reaction between the drug and the polymer at the high temperature.

Delivery of proteins is always challenging because these fabrication methods require either contact with organic solvent or high temperature. To minimize protein denaturation, a photopolymerization or photocross-linking approach may be advantageous. The protein is dispersed in vinyl monomers or water soluble polymers. The drug delivery matrix can then be formed by photoreaction at low temperatures. Naturally any reaction between the protein and the activated growing polymer chain or cross-linker should be minimal. If the drug delivery device is in the form of microsphere or microcapsule, different fabrication techniques are available. Microencapsulation is a mature and active field. Depending on the physicochemical properties of the drug and the polymer, drug-loaded microspheres can be prepared by solvent evaporation, solvent removal, hot-melt precipitation, interfacial polymerization, fluidized bed coating, simple and complex coacervation, polymer-polymer immiscibility induced phase separation, etc (Deasy, 1984; Karsa and Stephenson, 1993). Some of these processes, for example, complex coacervation, can be done at low temperatures and in aqueous solutions, rendering them ideal for protein encapsulation (Shao and Leong, 1995; Truong et al, 1995a).

Unique to tissue engineering applications is that the drug delivery matrix may need to be macroporous. Cell ingrowth and implant vascularization are appealing, as in cartilage and bone regeneration (Hollinger, 1993; Vacanti and Vacanti, 1994). Pore sizes above 150 μm would be desirable. The common method of fabricating a macroporous structure, or a foam, is to mix porogens with the polymer in the molding process, followed by an aqueous washout step to remove the porogens (Crane et al, 1995; Mooney et al, 1994). Other techniques create a porous structure by weaving or meshing together the polymer fibers (Mikos et al, 1993). However, these fabrication techniques are not suitable for loading growth factors because of the need to leach out the porogens. Recently, a foam fabrication process that allows the incorporation of a drug delivery function has been developed based on the principle of phase separation from a homogeneous solutions (Kadiyala et al, 1994; Lo et al, 1995). In essence, the polymer is dissolved in a solvent, and then the solution temperature is lowered until liquid-liquid phase separation is induced. This phase separation region is called the spinodal region where the two distinct liquid phases, a polymer phase and a solvent phase, emerge. The size of the phase domains depends on the temperature and the duration at which the solution is maintained at that temperature. The closer the temperature to the homogenous region and shorter the duration, the finer is the phase structure obtained. The morphology at any stage can be preserved by rapid quenching of the solution to prevent any coarsening of the microstructure. When the solvent is subsequently removed without disturbing this morphology, a highly porous structure will remain. The key to success is that the solvent can be removed without aqueous washing, as that would leach out any imbedded bioactive agents. This can be accomplished by using a solvent that can be removed by sublimation. Naphthalene, which is a good solvent for many biodegradable polymers used for tissue engineering, has a melting point of around 80°C and a low heat of vaporization for removal by sublimation, has been used.

With this technique foams made of poly(lactides) (PLLA) and poly(phosphoesters) with pore sizes ranging from 20–500 µm were fabricated (Kadiyala et al, 1996). Total pore surface area could be up to 1.3 m^2 per gram of foam, and total pore volume was nearly 18 cm^3/g. BMP-7 was incorporated into a PLLA foam and released in a near constant rate of 1.78 ng/hr for over two months after an initial burst. Bioactivity of the released BMP was confirmed in an ectopic bone formation assay. At 11 days post-implantation of these BMP-loaded foams in the subcutaneous space or rats, presence of chondrocytes was clearly evident inside the foams. By day 21, bone formation in the foam was manifested histologically and indicated by a high alkaline phosphatase level. Osteoblast lining and bone marrow elements were also observed in some samples.

This fabrication technique can be applied to any biodegradable polymer as long as the polymer is soluble in a sublimable solvent. High temperature is still required, and the protein has to be thermally robust. The release kinetics is sensitive to the distribution of the drug in the polymer phase. Another strategy to incorporate a controlled release function into a foam is to separately prepare the controlled release microspheres, and then distribute the microspheres in the foam. This approach eliminates the high temperature requirement and provides an extra dimension to control the release profile, provided uniform distribution and retention of the micro-spheres in the foam can be achieved, which may be difficult.

CELLULAR DRUG DELIVERY

The most elegant form of drug delivery is cellular. The complex and sophis-ticated cellular machinery, which can deliver the desired biomolecule in response to a stimulus in a tightly regulated control manner, is beyond match for a synthetic delivery device. An interesting drug delivery approach is to genetically engineer easily transfectable cell lines from nonautologous source to secrete a desired gene product. This approach was demonstrated in the secretion of significant levels of human growth hormone for weeks from mouse fibroblasts implanted in rat thymus (Behara et al, 1992; Doering and Chang, 1991). However, the novel gene product provoked an intense antibody response from its host recipient. Immunoprotecting the allogenic or xenogenic cells by a membrane permeable to small molecules and the desired gene product but not to lymphocytes and immunoglobulins repre-sents an intriguing approach to deliver proteins.

Methods of Cell Encapsulation

The use of microencapsulated cells as hybrid artificial organs was first pro-posed in 1964. Endocrine, islets, and hepatocytes were proposed to be encap-sulated by microspheres formed by the complexation between alginate and calcium. Intensive studies of these artificial cells, however, began only in the last decade; earlier studies failed to produce semipermeable microcapsules that have the right permeability and soft tissue biocompatibility.

Gelation of alginates is the most extensively studied system in cell encap-sulation (Chang, 1972; Chang, 1992). Alginate is a glycuranan extracted from brown seaweed algae. Calcium or other multivalent counterions chelates contiguous blocks of α-1,4-L-guluronan residues present in the polysaccha-ride. Cell encapsulation is achieved when cells suspended in an alginate solution are dropped or extruded into a solution containing calcium ions. The microcapsules formed can further be coated by adsorption of polyions such as polylysine, which can be coated by alginate again. Many cell types, including islets, hepatocytes, PC12 cells, chondrocytes, and fibroblasts, have been encapsulated by this method.

The advantage of the complex coacervation technique is that no organic solvent is involved. Other acidic polysaccharides, such as modified cellulose, can also be used. Amenable to both natural and synthetic systems, this technique is simple and yet versatile. The disadvantage is that the mechani-cal and chemical stability of the chelated biopolymer may be low. Totally synthetic membranes can be more stable, but they might not be the optimal substrates for cell growth and function.

Another technique of encapsulating cells is by interfacial precipitation (Wells et al, 1993). Cell suspension and polymer solution are extruded separately through two concentrically configured needles into a precipitating bath. Organic solvents such as dimethyl sulfoxide (DMSO), dimethyl formamide (DMF), dimethyl acetamide (DMAc), diethyl phthalate, acetone, are used to dissolve the organic polymers. Contact of cells with organic solvents is unavoidable but can be minimized through various coextrusion schemes. Encapsulation of chromaffin and PC12 cells was achieved by this technique using poly(acrylonitrile-co vinyl chloride) as the membrane, in configurations of 1 cm long hollow fibers (Aebischer et al, 1991a,b,c). *In vitro* and *in vivo* studies showed that both cell viability and cell functions were largely preserved in the encapsulation process. EUDRAGIT RL, a water-insoluble polyacrylate, has also been used as the membrane for encapsulating erythrocytes and fibroblasts (Boag and Sefton, 1987). Diethyl phthalate was used as the organic solvent, and a mixture of corn oil and mineral oil was used as the precipitating bath. Fibroblasts did not grow in the microcapsules unless collagen was also coencapsulated. Presumably this polyacrylate is not optimal for anchorage-dependent cells. The EUDRAGIT RL microcapsules are also fragile (Broughton and Sefton, 1989). Subsequently, the same researchers used cationic polyacrylates involving the copolymers of MMA-DMAEMA to improve the mechanical stability (Mallabone et al, 1989). Continued improvement has led to macroporous MMA-hydroxyethyl methacrylate (MMA-HEMA) microcapsules that have higher permeability (Crooks et al, 1990). Coencapsulating Matrigel, a reconstituted extracellular matrix derived from mouse tumor basement membrane, did improve the cell viability drastically.

The various synthetic systems demonstrate the degree of freedom in designing the semipermeable membrane. The microcapsules are mechanically and chemically more stable than the polyelectrolyte gels composed of polysaccharides. However, low permeability may be an issue. These synthetic polymers are also not optimal substrates for cell attachment, growth, and functions. Recently an approach to combine the strengths of the natural and synthetic systems was proposed (Shao et al, 1994a,b). The microcapsules were synthesized by a complex coacervation technique, composed of a combination of the following components: collagen or poly(dimethyaminoethyl methacrylate) as the polycation, hyaluronic acid or poly(methacrylic acid) as the polyanion, and poly(2-hydroxyethyl methacrylate) as an adjuster of the hydrophilicity of the microcapsules. In all cases, the inside layer of the microcapsules was composed of either collagen or hyaluronic acid; popular substrates for attachment and maintenance of cell functions of many cell types. The external layer was composed of synthetic polyelectrolytes, which imparted stability and allowed fine-tuning of the transport properties. Microcapsules encapsulating fibroblasts and cytokine gene transduced melanoma cells have been studied. The microcapsules remained stable in phosphate buffer at 37°C for months. Diffusion experiments conducted in diffusion cell showed that the membrane was permeable to BSA (MW = 67,000) but not to alcohol dehydrogenase (MW = 150,000). Implantation studies in mice demonstrated that the encapsulated cells could maintain their viability and functions for at least three weeks.

Drug Delivery by Encapsulated Cells

An exciting application of encapsulated cells is an artificial pancreas. Islets, isolated from pigs and encapsulated by a semipermeable membrane for immunoprotection can secrete insulin in response to glucose level for diabetic therapy. Although long term viability of the encapsulated islets and fibrous tissue reaction to the capsules are barriers to be overcome, excellent glycemic control without immunosuppression has been observed in different animal models (Maki et al, 1995). Success of this cellular approach in providing a closed-loop control would alleviate many undesirable effects associated with chronic irregular glucose levels.

Intracranial implantation of polymer encapsulated neurotransmitter secreting cells has been proposed for various central nervous system (CNS) deficits (Aebischer et al, 1991c; Emerich et al, 1992; Langer and Vacanti, 1993; Tresco et al, 1992). This is an attractive idea since drug delivery to the brain is always plagued by low bioavailability caused by the presence of the blood-brain-barrier (BBB). As only small or lipophilic agents can cross the BBB, potent biomacromolecules such as nerve growth factors cannot be effectively delivered by conventional means. In response to dopamine deficiency associated with Parkinson's disease, local drug delivery systems such as implantable osmotic pumps or controlled release polymers have been implanted intracranially to remedy the deficit (Becker et al, 1990). While improved function has been reported in various experimental animal models, there are problems of dopamine autooxidation and cavitation around the injection site. As frequent replacements of devices in the brain is highly impractical, the limited service lifetime of controlled release systems also renders this approach less attractive. Transplantation of dopaminergic tissues into the striatum represents a potential solution. Nevertheless, although the transplanted fetal neurons can survive and make synaptic contacts with the host striatal neurons, there are formidable hurdles to be overcome. They include failure to reestablish the normal neural circuitry, high mortality and morbidity associated with the transplant procedure, and the ethical issue of human fetal tissue research. It is also believed that over time the transplanted tissue will be rejected even if allogeneic tissue is used. To circumvent some of these obstacles, bovine adrenal medullary chromaffin cells and

PC12 cells have been encapsulated in alginate-polylysine microcapsules or poly(acrylonitrile vinyl chloride) hollow fibers (Aebischer et al, 1991b; Tresco et al, 1992). *In vitro* studies showed that at least some of the cells survived the encapsulation procedure. Release of dopamine from both the microcapsules and macrocapsules was observed in response to a chemically-induced depolarization. Encapsulated PC12 cells alleviated lesion-induced rotational asymmetry in rats for least four weeks (Aebischer et al, 1991a). Immunoprotection was demonstrated when both types of microcapsules were implanted in an immunologically incompatible host. Implanted in *Macaca fascicularis* monkeys that exhibited hemiparkinsonian symptoms, these polymer-encapsulated PC12 cells also induced significant behavior improvement in the animals (Aebischer et al, 1994). A recent study showed that this approach might find applications in human motoneuron diseases such as amyotrophic lateral sclerosis (Sagot et al, 1995; Zurn et al, 1994). Polymer-encapsulated BHK cells transduced with the glial cell line-derived neurotrophic factor (GDNF) gene reduced the death of motoneurons in newborn rats with transected facial nerves.

Pain management is one of the major challenges of rehabilitation medicine. Conventional pharmacological intervention always requires escalating doses and repeat administration. A recently reported promising approach for chronic pain management is to transplant adrenal medullary chromaffin cells into the spinal subarachnoid space (Sagen, 1992). In rats the transplanted cells survived for months and released high levels of opioid peptides and catecholamines. In behavioral studies in rats, the transplants reduced pain in an arthritis model and a peripheral neuropathy model. Subsequent limited clinical trials demonstrated that the patients received pain relief over a period of 4–10 months and a concomitant decrease in narcotic intake. Increased levels of catecholamines and metencephalon in the spinal CSF samples of patients were also observed. Success of this clinical trial relied on the availability of human adrenal glands and the administration of the immunosuppressive agent cyclosporine A for two weeks. To explore the use of xenogenic cells, bovine chromaffin cells were encapsulated and implanted into the rat spinal subarachnoid space (Sagen et al, 1993). The encapsulated cells provided a sustained release of met-enkephalin and catecholamines and reduced pain sensitivity with nicotine stimulation in the animals for up to three months. It appears that microcapsulated cells can bolster immensely the appeals of this cell-based management of chronic and maybe even intractable pain.

GENE TRANSFER VECTORS FOR DRUG DELIVERY

A variation of cellular drug delivery is to directly transduce the cells *in vivo* to secrete the target protein. Research in this area is pursued intensely around the globe, driven by the immense interest in delivering exogenous genes for correcting genetic diseases. Gene therapy for cystic fibrosis (CF) is examined here to highlight the promise and challenges of protein expression by *in vivo* gene transfer.

Cystic fibrosis is one of the most approachable genetic diseases amenable to gene therapy. The molecular biology of CF and the CFTR gene is well studied and characterized (Cai et al, 1989; Drumm and Collins, 1993); the cellular target is accessible. Data suggest that clinical benefit can be observed at relatively low levels of expression, and only 6–10% of the airway epithelial cells need to be transfected for partial correction of the chloride transport defects. Impressive progress has been made since the discovery of the CFTR gene in 1989. Molecular evidence of gene transfer has been clearly observed in preclinical and clinical studies using liposomes (Caplen et al, 1995; Logan et al, 1995), adenovirus (Bout et al, 1994; Grubb et al, 1994; Knowles et al, 1995; Rich et al, 1993; Rosenfeld et al, 1992; Zabner et al, 1993), and adeno-associated virus vectors (Flotte et al, 1993a,b) to deliver the CFTR gene to lung airway epithelial cells. These studies demonstrated the feasibility of gene therapy for the pulmonary complications of CF. The most recent clinical studies also suggested, however, that the first generation vectors needed to be improved (Caplen et al, 1995; Knowles et al, 1995). The study with the liposomal vector indicated that the transfection efficiency and the duration of CFTR expression were therapeutically unsatisfactory. The trial using an adenoviral vector revealed that there was no significant normalization of the chloride and sodium transport and that there were signs of toxic effects evoked by high doses of the vector. Refinement of the gene delivery vectors is being intensively pursued by many groups. Several excellent reviews appeared recently on the relative advantages and limitations of the different vectors (Crystal, 1995a,b; Ledley, 1995; O'Neal and Beaudet, 1994).

Viral Gene Delivery Vectors

Currently, the majority of approved gene therapy trials in humans relies on viral vectors for gene transduction. Replication-defective retroviral vectors achieve high efficiency of gene transfer and stable integration of the gene into cellular DNA in mitotically active cells. Adenovirus in particular is an attractive candidate for gene therapy of cystic fibrosis because of its natural tropism for respiratory tract cells and ability to infect nonreplicating cells. The efficacy of this viral system in humans is supported by studies showing that E1-deficient adenovirus encoding CFTR administered to the nasal airway epithelium of three patients with CF corrected the characteristic Cl⁻ transport defect in the transfected cells (Zabner et al, 1993), although another clinical trial cannot confirm this functional correction (Knowles et al,

1995). Adenovirus DNA rarely integrates; thus repeated administration will be required for treatment of CF, raising concerns about the safety and efficacy of this vector (Crystal 1995a,b; Yang et al, 1995). With repeated administration of the adenovirus-CFTR vector applied to intrapulmonary airway epithelia of cotton rats and nasal epithelia of Rhesus monkeys, an antibody response but no inflammatory response was detected (Zabner et al, 1994).

The adeno-associated virus (AAV) is also an attractive vector because it is nonpathogenic, can be devoid of viral coding sequences, and has the potential for high frequency stable integration at a specific site on chromosome 19 mediated by the AAV Rep78 and Rep68 proteins (Giraud et al, 1994; Samulski et al, 1991; Weitzman et al, 1994). AAV-CFTR vectors have been shown to transfect bronchial epithelial cells with expression of the CFTR gene, both in cell culture and *in vivo* (Egan et al, 1992; Flotte et al, 1993a,b). Since the viral vectors must be prepared in cultured cells, there exists the possibility of contamination by microorganisms, nucleic acids, and replication-competent helper virus, which may lead to infection of replicating, brain, or other sensitive susceptible cells. The potential problem of insertional mutagenesis resulting from the integration of the new sequences into cellular DNA and the possible deleterious consequences of an immune response also remained to be investigated.

A limiting factor common to all viral vectors is the safety concern of repeated administration. The lifespan of airway epithelial cells in general is shorter than 120 days. Thus to stably transduce the airway epithelium for longer than a few months, it is necessary to stably transduce a precursor cell population. Gene delivery to a precursor cell(s) has not yet been proven. Repeated treatment with viral vectors may be problematic due to the development of an immune response against the viral proteins. Problems may also occur with patients who have previously been infected with adenovirus. These possible complications have stimulated intensifying research effort to develop nonviral gene delivery systems.

Nonviral Gene Delivery Vectors

Although synthetic transfection systems can result in only transient gene expression, they have the advantage of using purified DNA. Another advantage of the nonviral system over the viral vectors is that plasmids of any size can be used, compared to a limit of 7–10 kilobases for the DNA to be packaged into a virus. The negatively charged DNA can be complexed with cationic lipid or polylysine-ligand to form soluble aggregates that can be endocytosed (Cotten and Wagner, 1993; Felgner and Rhodes, 1991; Felgner et al, 1994; Ferkol et al, 1993; Ferkol et al, 1995; Ledley, 1995; Logan et al, 1995; Schofield and Caskey, 1995; Singhal and Huang, 1994; Wagner et al,

1990; Wagner et al, 1992; Wagner et al, 1993; Wu and Wu, 1988). Those that incorporate specific ligands have the potential of targeted gene delivery through receptor-mediated endocytosis, as has been demonstrated with a large number of ligand-DNA conjugates. The liposome technology has been applied to cystic fibrosis and tested by tracheal installation in transgenic (cf/cf) mice. CFTR expression was found in epithelial cells of both large and small airways, demonstrating the potential for this type of delivery vehicle (Hyde et al, 1993). Coupled to mechanisms for the disruption of endosomes, i.e., by use of polylysine conjugated to a replication-defective adenovirus as an endosomal lysis vector (Cristiano et al, 1993a,b; Curiel et al, 1991; Gao et al, 1993), or endosome-disruptive peptides (Wagner et al, 1992; Plank et al, 1994), these systems have shown high level expression of marker genes *in vitro*. While the conjugation of adenovirus to the complex resulted in gene expression levels 100-fold greater than those observed in a similar protocol using cationic liposomes, the potential immunologic response to the adenovirus proteins may limit repetitive dosing. A recent clinical trial using the cationic liposome DC-Chol/DOPE (3b[N-(N′,N′-dimethylaminoethane)-carbamoyl)-cholesterol/dioleolyphosphatidylethanolamine) to deliver the CFTR cDNA to the nasal epithelia of CF patients showed molecular evidence of transfection (Caplen et al, disappearing by day seven, was observed in some patients. It was concluded, however, that the efficiency of transfection and duration of gene expression of this gene delivery system would need to be improved to be therapeutically effective. The success of the liposome-DNA and polylysine-ligand-DNA approaches appear to be limited by two factors. The first is the instability of the ionic complexation. Held together only by electrostatic interaction, these complexes often dissociate in biological fluids because of the screening effect of the natural polyelectrolytes in the fluids. For instance, while aggregates of transferrin-polylysine-DNA transfect many different cell types efficiently *in vitro,* they are ineffective *in vivo* (Service, 1995). The second factor is the susceptibility of the cDNA to nuclease degradation in transit from site of administration to the nucleus of the target cell. A recent pharmacokinetic study showed that a HLA-B7 gene, when delivered by DC-Chol/DOPE intravenously, had a half-life of less than five minutes, and became undetectable after one hour (Law et al, 1995). No encoded protein was detected in tissues harboring the residual plasmid by immunohistochemical analysis 1 or 7 days post-injection. This poor bioavailability is not too surprising in light of similar instability characteristic of many therapeutic proteins. Recently a gene delivery system based on the complex coacervation of gelatin and DNA was proposed (Truong et al, 1994; Truong et al, 1995b).

The gelatin-DNA coacervate has elements in common with the cationic lipid-DNA condensates and ligand-polycation-DNA complexes in the physical delivery of DNA to the cell through endocytosis, but differs in the aspect that the cDNA is physically entrapped in a solid nanoparticle and the coacervate is stable extracellularly. Targeting ligands can be conjugated to the surface of the DNA-nanoparticles through the amino side chains of gelatin. To improve the cellular uptake of these nanoparticles, targeting ligands such as transferrin can be covalently attached to the surface of the nanosphere for receptor-mediated endocytosis. Another advantage of this gene delivery system is that endolysosomolytic agents can be coencapsulated and released intracellularly to promote escape of intact DNA into the cytoplasm. A major barrier to efficient transfection is the degradation of the DNA in the endolysosomal compartments, where the nanoparticles would be sequestered after phagocytosis. Chloroquine, which blocks the pathway between the endosomal and lysosomal compartments, has been incorporated into the DNA-nanospheres to improve the transfection efficiency. Evaluated in the Luciferase gene-293s cell system, the gene transfer level, 10^{8-9} U/mg, as achieved by the DNA-nanospheres was within range of that observed with the liposomal vector (10^{9-10} U/mg) and the $CaPO_4$-DNA coprecipitation technique ($\approx 10^{10}$ U/mg). The *in vitro* efficacy of this gene delivery system has been extended to other genes (LAMP-1, *LacZ,* CFTR, and NOS) and other cell lines (U937, HeLa, 3HTE, CHO, and COS7). The potential of the gelatin-DNA nanospheres serving as an *in vivo* gene therapy vector for cystic fibrosis was tested in New Zealand white rabbits. Nanospheres containing 347 µg of the pSA306 CFTR plasmid DNA were injected into the right lower lobes of adult rabbits using a pediatric bronchoscope with fiber optic camera. Reverse transcriptase PCR and in situ PCR were performed on tissue homogenates from the right lower lobe of the lung seven days post-transfection. A positive signal for the presence of the exogenous gene in the epithelial cells lining the submucosal layer of the lung airway was clearly observed in animals given the DNA nanospheres. Control animals which received the naked DNA at the same dose showed no signs of transfection.

CONCLUDING REMARKS

Polymeric controlled drug delivery has made a significant impact on drug therapy involving genetically engineered polypeptides. It will also make important contributions to tissue engineering. Cell growth, differentiation, and tissue morphogenesis are undoubtedly influenced by the extracellular matrix as well as regulated by soluble growth factors. As the mystery of these cellular processes becomes unraveled, strategic application of controlled delivery will effect or accelerate the formation of the desired tissue. Cellular approaches, which can be more elegant, will continue to augment the polymeric mode of drug delivery. The goal of *in vivo* gene transfer for drug delivery is more reachable than that for amelioration of genetic defects. The latter requires persistent correction, while drug delivery may need only be transient. The technology for transient *in vivo* transfection already exists; the issues to be addressed are product reproducibility and long-term biocompatibility. The future of delivery of polypeptides can be expected to be significantly bolstered by the gene transfer approach, either directly *in vivo* or with immunoprotection.

References

Aebischer P, Tresco PA, et al. (1991a): Transplantation of microencapsulated bovine chromaffin cells reduces lesion-induced rotational asymmetry in rats. *Brain Res* 560: 43

Aebischer P, Wahlberg L, et al. (1991b): Macroencapsulation of dopamine-secreting cells by coextrusion with an organic polymer solution. *Biomaterials* 12: 50

Aebischer P, Winn SR, et al. (1991c): Transplantation of polymer encapsulated neurotransmitter secreting cells: Effect of the encapsulating technique. *Trans ASME* 113: 178

Aebischer P, Goddard M, et al (1994): Functional recovery in hemiparkinsonian primates transplanted with polymer-encapsulated PC12 cells. *Exp Neurol* 126: 151

Albin G, Horbett TA, et al. (1986): *Advances in Drug Delivery Systems*. Amsterdam: Elsevier

Becker J, Robinson TE, et al. (1990): Sustained behavioral recovery from unilateral nigrostriatal damage produced by the controlled release of dopamine from a silicone polymer pellet placed into the denervated striatum. *Brain Res* 508: 60

Behara AM, Westcott AJ, et al. (1992): Intrathymic implants of genetically modified fibroblasts. *FASEB J* 6: 2853

Boag AH, Sefton MV (1987): Microencapsulation of human fibroblasts in a water-insoluble polyacrylate. *Biotech Bioeng* 30: 954

Bout A, Perricaudet M, et al. (1994): Lung gene therapy: In vivo adenovirus-mediated gene transfer to rhesus monkey airway epithelium. *Human Gene Therapy* 5: 3

Broughton RL, Sefton MV (1989): Effect of capsule permeability on growth of CHO cells in Eudragit RL microcapsules: Use of FITC-dextran as a marker of capsule quality. *Biomaterials* 10: 462

Brunelli GA, Vigasio A, et al. (1994): Different conduits in peripheral nerve surgery. *Microsurgery* 15: 176

Cai Z, Shi Z, et al. (1989): Development and evaluation of a system of microencapsulation of primary rat hepatocytes. *Hepatology* 10: 855

Caplen NJ, Altons E, et al. (1995): Liposome-mediated CFTR gene transfer to the nasal epithelium of patients with cystic fibrosis. *Nature Medicine* 1: 39

Chang TMS (1972): *Artificial Cells*. Springfield, IL: Charles C. Thomas

Chang TMS (1992): Hybrid artificial cells: Microencapsulation of living cells. *ASAIO J* 38: 128

Cieslinski DA, Humes HD (1994): Tissue engineering of a bioartificial kidney. *Biotech Bioeng* 43: 678

Cotten M, Wagner E (1993): Non-viral approaches to gene therapy. *Curr Opin Biotech* 4: 705–710

Crane GM, Ishaug SL, et al. (1995): Bone tissue engineering. *Nature Medicine* 1: 1322

Cristiano R, Smith L, et al. (1993a): Hepatic gene therapy: Efficient gene delivery and expression in primary hepatocytes utilizing a conjugated adenovirus DNA complex. *Proc Natl Acad Sci USA* 90: 11548

Cristiano R, Smith L, et al. (1993b): Hepatic gene therapy: Adenovirus enhancement of receptor-mediated gene delivery and expression in primary hepatocytes. *Proc Natl Acad Sci USA* 90: 2122.

Crooks CA, Douglas JA, et al. (1990): Microencapsulation of mammalian cells in a HEMA-MMA copolymer: Effects on capsule morphology and permeability. *J Biomed Matr Res* 24: 1241

Crystal RG (1995a): The gene as the drug. *Nature Medicine* 1: 15

Crystal RG (1995b): Transfer of genes to humans: Early lessons and obstacles to success. *Science* 270: 404

Curiel D, Agarwal S, et al. (1991): Adenovirus enhancement of transferrin-polylysine-mediated gene delivery. *Biochemistry* 88: 8850

Deasy P (1984): *Microencapsulation and Related Drug Processes.* New York: Marcel Dekker

Doering LC, Chang PL (1991): Expression of a novel gene product by transplants of genetically modified primary fibroblasts in the central nervous system. *J NeuroSci Res* 29: 292

Drumm ML, Collins FS (1993): Molecular biology of cystic fibrosis. *Mol Gen Med* 3: 33

Egan M, Flotte T, et al. (1992): Defective regulation of outwardly rectifying CI channels by protein kinase A corrected by insertion of CFTR *Nature* 358: 581

Emerich DF, Winn SR, et al. (1992): A novel approach to neural transplantation in Parkinson's disease: Use of polymer-encapsulated cell therapy. *Neurosci Biobehav Rev* 16: 437

Fan LT (1989): *Controlled Release: A Quantitative Treatment.* New York: Springer-Verlag

Felgner PL, Rhodes G (1991): Gene therapeutics. *Nature* 349: 351

Felgner JH, Kumar R, et al. (1994): enhanced gene delivery and mechanism studies with a novel series of cationic lipid formulations. *J Biol Chem* 269: 2550

Ferkol T, Kaetzel CS, et al. (1993): Gene transfer into respiratory epithelial cells by targeting the polymeric immunoglobulin receptor. *J Clin Invest* 92: 2394

Ferkol T, Perales JC, et al. (1995): Gene transfer into the airway epithelium of animals by targeting the polymeric immunoglobulin receptor. *J Clin Invest* 95: 493

Flotte T, Afione S, et al. (1993a): Stable in vivo expression of the cystic fibrosis transmembrane conductance regulator with an adeno-associated virus vector. *Proc Natl Acad Sci USA* 90: 10613

Flotte T, Afione S, et al. (1993b): Expression of the cystic fibrosis transmembrane conductance regulator from a novel adeno-associated virus promotor. *J Biol Chem* 268: 153781

Gao L, Wagner E, et al. (1993): Direct in vivo gene transfer to airway epithelium employing adenovirus-polylysine-DNA Complexes. *Human Gene Ther* 4: 17–24

Giraud C, Winocour E, et al. (1994): Site-specific integration by adeno-associated virus is directed by a cellular DNA sequence. *Proc Natl Acad Sci USA* 91: 10039

Greco RS (1994): *Implantation Biology: The Host Response and Biomedical Devices.* Boca Raton: CRC Press

Grubb BR, Pickles RJ, et al. (1994): Inefficient gene transfer by adenovirus vector to cystic fibrosis airway epithelia of mice and humans. *Nature* 371: 802

Hollinger J (1993): Factors for osseous repair and delivery: Part 11. *J Craniofac Surg* 4: 135

Hyde S, Gill D, et al. (1993): Correction of the ion transport defect in cystic fibrosis transgenic mice by gene therapy. *Nature* 362: 250

Jeong SY, Kim SW, et al. (1986): Self-regulating insulin delivery systems III. In vivo studies. In: Kim SW, Anderson J, eds. *Advances in Drug Delivery Systems.* Amsterdam: Elsevier

Kadiyala S, Lo H, et al. (1994): Biodegradable polymers as synthetic bone grafts. In: Brighton CT, Friedlaender G, Lane J, eds. *Bone Formation and Repair.* Rosemont, Illinois: American Academy of Orthopedic Surgeons

Kadiyala S, Lo H, et al. (1996): *Bone Induction Achieved by Controlled Release of BMP From PLA/Hydroxyapatite Foams.* Fifth World Biomaterials Congress Toronto: ???

Karsa DR, Stephenson RA (1993): *Encapsulation and Controlled Release.* Cambridge, England: Royal Society of Chemistry

Knowles MR, Hohneker KW, et al. (1995): A controlled study of adenoviral-vector-mediated gene transfer in the nasal epithelial of patients with cystic fibrosis. *NEJM* 333: 823

Langer R (1990): New methods of drug delivery. *Science* 249: 1523

Langer R, Vacanti JP (1993): Tissue engineering. *Science* 260: 920

Langer R, Wise D (1986): *Medical Applications of Controlled Release.* Boca Raton: CRC

Law D, Parker S, et al. (1995): Cancer gene therapy using plasmid DNA: Pharmacokinetic study of DNA following injection in mice. *Hum Gene Ther* 6: 553

Ledley FD (1995): Nonviral gene therapy: The promise of genes as pharmaceutical products. *Hum Gene Ther* 6: 1129

Leong KL, Langer R (1987): Polymeric controlled drug delivery. *Adv Drug Del Rev* 1: 199

Lo H, Ponticiello M, et al. (1995): Fabrication of controlled release biodegradable foams by phase separation. *Tissue Eng* 1: 15

Logan JJ, Bebok Z, et al. (1995): Cationic lipids for reporter gene and CFTR transfer to rat pulmonary epithelium. *Gene Ther* 2: 38

Maki T, Mullon CJ, et al. (1995): Novel delivery of pancreatic islet cells to treat insulin-dependent diabetes mellitus. *Clin Pharmacokinet* 28: 471

Mallabone CL, Crooks CA, et al. (1989): Microencapsulation of human diploid fibroblasts in cationic polyacrylates. *Biomaterials* 10: 380

Mikos AG, Bao Y, et al. (1993): Preparation of poly-(glycolic acid) bonded fiber structures for cell attachment and transplantation. *J Biomed Matr Res* 27: 183

Mooney DJ, Kaufmann PM, et al. (1994): Transplantation of hepatocytes using porous, biodegradable sponges. *Transplant Proceed* 26: 3425

Nash J, Pini A (1996): Making the connections in nerve regeneration. *Nature Medicine* 2: 25

O'Neal WK, Beaudet AL (1994): Somatic gene therapy for cystic fibrosis. *Hum Mol Gen* 3: 1497

Peppas NA, Langer R (1994): New challenges in biomaterials. *Science* 263: 1715

Plank C, Oberhauser B, et al. (1994): The influence of endosome-disruptive peptides on gene transfer using synthetic virus-like gene transfer systems. *J Biol Chem* 17: 12918

Reddi AH (1994): Symbiosis of biotechnology and biomaterials: Applications in tissue engineering of bone and cartilage. *J Cell Biochem* 56: 192

Rich D, Couture L, et al. (1993): Development and analysis of recombinant adenoviruses for gene therapy of cystic fibrosis. *Hum Gene Ther* 4: 461

Rosenfeld M, Yoshimura K, et al. (1992): In vivo transfer of the human cystic fibrosis transmembrane conductance regulator gene to the airway epithelium. *Cell* 68: 143

Sagen J (1992): Chromaffin cell transplants for alleviation of chronic pain. *ASAIO J* 38: 24

Sagen J, Wang H, et al. (1993): Transplants of immunologically isolated xenogeneic chromaffin cells provide a long-term source of pain-reducing neuroactive substances. *J Neurosci* 13: 2415

Sagot Y, Tan SA, et al. (1995): Polymer encapsulated cell lines genetically engineered to release ciliary neurotrophic factor can slow down progressive motor neuronopathy in the mouse. *Eur J Neurosci* 7: 1313

Samulski RJ, Zhu X, et al. (1991): Targeted integration of adeno-associated virus (AAV) into human chromosome 19. *EMBO J* 10: 3941

Schofield JP, Caskey CT (1995): Non-viral approaches to gene therapy. *Brit Med J* 51: 56

Service RF (1995): Dendrimers: Dream molecules approach real applications. *Science* 266: 458

Shao W, Leong KW (1995): Complex coacervation of chondroitin sulfate and gelatin: Microsphere formation and encapsulation of a model protein. *J Biomat Sci* 5: 389

Shao W, Gutsche A, et al. (1994a): Encapsulation of Cells by Complex Coacervation of Collagen and Synthetic Polyanions. *Trans Soc Biomater* 17: 316

Shao W, Jaffee E, et al. (1994b): *Delivery of Cytokines by Microencapsulated Transduced Cells*. Nice, France: International Conference on the Controlled Release of Bioactive Agents

Singhal A, Huang L (1994): Gene transfer in mammalian cells using liposomes as carriers. In: Wolf J, ed. *Gene Therapeutics: Methods and Applications of Direct Gene Transfer*. Boston: Birkhauser

Tresco PA, Winn SR, et al. (1992): Polymer encapsulated neurotransmitter secreting cells. *ASAIO J* 38: 17

Truong V, Guarnieri F, et al. (1994): *Immuno-microsphere as Gene Delivery Vehicle: Targeting of LAMP-1 to Lysosomal membrane*. Nice, France: Proceedings of the International Symposium on Controlled Release of Bioactive Materials

Truong V, Williams JR, et al. (1995a): Targeted delivery of immunomicrospheres in vivo. *Drug Del* 2: 166

Truong VL, Walsh SM, et al. (1995b): *Gene transfer by gelatin-DNA coacervate*. Seattle WA: Proceedings of the International Symposium on Controlled Release of Bioactive Materials

Vacanti CA, Vacanti JP (1994): Bone and cartilage reconstruction with tissue engineering approaches. *Otolaryngol Clin N Am* 27: 263

Wagner E, Zenke M, et al. (1990): Transferrin-polycation conjugates as carriers for DNA uptake into cells. *Proc Natl Acad Sci USA* 87: 3410

Wagner E, Plank C, et al. (1992): Influenza virus hemagglutinin HA-2 terminal fusogenic peptides augment gene transfer by transferrin-polylysine-DNA complexes: Toward a synthetic virus-like gene-transfer vehicle. *Proc Natl Acad Sci USA* 89: 7934

Wagner E, Curiel D, et al. (1993): Delivery of drugs, proteins and genes into cells using transferrin as a ligand for receptor-mediated endocytosis. *Adv Drug Del Rev* 14: 113

Weitzman M, Kyostio S, et al. (1994): Adeno-associated virus (AAV) rep proteins mediate complex formation between AAV DNA and its integration site in human DNA. *Proc Natl Acad Sci USA* 91: 5808

Wells GDM, Fisher MM, et al. (1993): Microencapsulation of viable hepatocytes in HEMA-MMA microcapsules: A preliminary study. *Biomaterials* 14: 615

Williams SK, Jarrell BE (1996): Tissue-engineered vascular grafts. *Nature Med* 2: 32

Wu GY, Wu CH (1988): Receptor-mediated gene delivery and expression in vivo. *J Biol Chem* 263: 14621

Xu XM, Guenard V, et al. (1995): A combination of BDNF and NT-3 promotes supraspinal axonal regeneration into Schwann cell grafts in adult rat thoracic spinal cord. *Exp Neurol* 134: 261

Yang Y, Li Q, et al. (1995): Cellular and humoral immune responses to viral antigens create barriers to lung-directed gene therapy with recombinant adenoviruses. *J Virol* 69: 2004

Zabner J, Couture L, et al. (1993): Adenovirus-mediated gene transfer transiently corrects the chloride transport defect in nasal epithelia of patients with cystic fibrosis. *Cell* 75: 207

Zabner J, Peterson D, et al. (1994): Safety and efficacy of repetitive adenovirus-mediated transfer of CFTR cDNA to airway epithelia of primates and cotton rats. *Nature Gen* 6: 75

Zurn AD, Baetge E, et al. (1994): Glial cell line-derived neurotrophic factor (GDNF), a new neurotrophic factor for motoneurones. *Neuroreport* 6: 113

7

SYNTHETIC BIODEGRADABLE POLYMER SCAFFOLDS

GAIL K. NAUGHTON, PH.D., RONNDA BARTEL, PH.D.,
JONATHAN MANSBRIDGE, PH.D.

INTRODUCTION

Over the last several decades, new technologies have enabled the in vitro cultivation of many different human cell types. At the same time, the art and science of medicine has been expanded to include new biopharmaceutical and recombinant genetic therapies. Organ transplantation has become commonplace, limited not by surgical technique but by donor availability. Tissue engineered organs can obviate the torturous waiting and the human tragedy caused by the scarcity of donor organs. Furthermore, tissue engineered organs can be used in testing procedures, reducing or eliminating the need for animal and human subject tests. The first tissue engineered organ, which has progressed from the lab bench to the first accepted patient care, has been skin, the body's largest organ (Rheinwald, 1989). Tissue engineering offers many benefits to the patient and has the potential of redefining transplantation. Researchers can use standard, well-recognized cell-banking procedures to characterize the human cell source for these tissue products and to safety-test this source for possible contaminants (US FDA; 1993a). Normal human tissues may be more efficacious than animal or synthetic tissues because normal human tissues supply a balanced natural mixture of growth factors and both structural and interactive functional proteins. These tissues can communicate with the body's own repair mechanisms to stimulate angiogenesis, remodeling, and other steps to restore complete function. Tissue engineered products can be stored in a frozen state, providing "off-the-shelf" availability for physicians, who previously have been restricted to

Synthetic Biodegradable Polymer Scaffolds
Anthony Atala and David Mooney, Editors
Robert Langer & Joseph P. Vacanti, Associate Editors
© 1997 Birkhäuser Boston

using the tissue and organs from the limited cadaveric and/or surgical discard sources.

Traditional culture of nontransformed animal and human cells involves the attachment and growth of the cells into two-dimensional flat sheets in containers such as petri dishes, flasks, and roller bottles. These monolayer cultures do not reflect the normal cellular environment; moreover, they potentiate artificial effects such as dedifferentiation (Lucas-Clerc et al, 1993). The core tissue engineering technology developed in our laboratory creates a *three-dimensional* living stroma and resultant tissue in vitro. This three-dimensional matrix culture allows cells to develop and assemble into tissues that closely resemble their in vivo counterparts (Contard et al, 1993).

In normal growth and development, specialized connective tissue cells form stroma, and this living framework provides the three-dimensional structure for each tissue and organ type. In addition to structural components, the stroma contains various attachment and growth factors that are required for full differentiation and function of the tissue. For each tissue engineering application, specific cells are seeded onto a polymer mesh framework in an optimum environment to stimulate differentiated tissue development. The cells attach, divide, and secrete extracellular matrix proteins and growth factors, using the hydrolyzable polymer mesh as a scaffolding (Langer and Vacanti, 1993). The resultant human tissue construct may then be transplanted to treat human medical applications (Hansbrough, 1995; Hansbrough and Morgan, 1993) or used in an in vitro model to replace and improve toxicology testing currently performed on animals or human subjects (Mayer et al, 1994; Perkins and Osborne, 1993).

The first tissue engineered products to move into commercial application were those based on the skin and epithelia as cell sources and the targets for therapeutic intervention. The largest organ in the human body, the skin provides the separation and barrier between the organism and the environment. It is a complex, highly organized structure of a continually proliferating epidermis, consisting of several cell types including keratinocytes which are responsible for differentiating into the stratum corneum. Fibroblasts, the main cellular component of the dermis, secrete various macromolecules including collagen, elastin, fibronectin, decorin, tenascin, laminin, and proteoglycans to form the connective tissue underlayment that provides the skin with its high tensile strength (DeWever and Rheins, 1994; Landeen et al, 1992).

When the skin is damaged or breaks down because of environmental insult or disease, the results can be minor or life-threatening. Therefore, the use of skin models to measure the level of environmental toxic and damaging factors such as chemicals or solar radiation in vitro have been the subject of tissue engineering research. Using tissue engineering, researchers have developed an in vitro skin model that may replace many toxicological tests currently performed on animals and humans, and this procedure has received its initial regulatory approval.

The therapeutic need for a tissue-engineered skin is great and several innovative type of tissue transplants are being studied for the treatment of burns, skin ulcers, and various deep wounds. Clinical studies have involved the use of composite materials to induce the formation of a neodermis in the *in vivo* application of *in vitro* culture of keratinocytes (Rheinwald, 1989; Yannas et al, 1982), cryopreserved allografts (Cuono et al, 1986), bovine collagen-based full-thickness skin equivalents (Bell et al, 1979), and bioengineered dermis (Economou et al, 1995; Hansbrough et al, 1992a,b; Takeda et al, 1995). The majority of experience in our laboratories involves the development and use of dermal-based tissue engineered products for the treatment of severe burns and diabetic ulcers. As such, this approach will be highlighted as an example of the development and use of a bioengineered skin.

TISSUE ENGINEERED DERMIS FOR THERAPEUTIC INTERVENTION

For medical applications, the dermal or lower layer of the skin has significant advantages that makes it suitable as an initial target for tissue engineered therapeutics. The human diploid fibroblast cell type, its main constituent, grows well in culture; it has a proliferative potential of approximately 60 population doublings (10^{18}-fold increase over the original cell isolation) and has long been used for the production of human biologicals (Kruse and Patterson, 1973; Petricciani and Hennessen, 1987; Petricciani et al, 1979). Unlike keratinocytes, which carry surface human leukocyte antigens (i.e., HLA-DR) that may cause allograft rejection phenomina (Hansbrough, 1990), implantation of allogenic human fibroblasts does not stimulate an immune response (Cuono et al, 1986; 1987). In clinical trials involving more than 400 patients, Advanced Tissue Sciences has implanted Dermagraft™ products of cultured human foreskin fibroblasts from two individual neonatal donors; no clinical evidence of immune response or rejection has been seen in any of these patients.

There is a major medical need for an effective dermal replacement because dermal tissue does not regenerate into normal dermis in vivo after serious burns; instead, dermal tissue potentiates the formation of scar tissue. "Loss of the dermis in extensive full-thickness burn wounds therefore poses a serious problem which is not completely solved by the application of meshed, split-thickness autograft skin" (Hansbrough, 1990). The meshing of the autograft in treatment of these burned patients allows the available unburned skin to provide the maximum coverage in as short a time as possible and is a life-saving procedure. The epidermis quickly spreads to fill in the mesh openings for "graft take" from 85% to 100%. In the long term, however, the lack of normal underlying vascularized dermis can result in skin fragility, blistering, and formation of hypertrophic scar. Animal experiments as well as clinical results have demonstrated this dependence on dermal integrity (Demarchez et al, 1992; Hansbrough, 1990, 1995; Hansbrough et al, 1993; Leary et al, 1992). This natural failure of the dermis to regenerate may also play a role in other wound situations, which have shown clinical response to dermal replacement as a therapeutic approach.

A TISSUE ENGINEERED DERMAL LAYER

Dermagraft has been designed to replace this dermal layer and potentiate a normal wound-healing process. Characterization studies have shown that Dermagraft is potentially a good model to study in vitro epithelialization (Landeen et al, 1992) and the interaction between the extracellular matrix and the adjacent epithelia (Contard et al, 1993). In preclinical studies and in pilot and pivotal human clinical trials, Dermagraft products have been evaluated as a dermal replacement under autograft for treating severe burn wounds (Economou et al, 1995; Hansbrough et al, 1992a,b; Takeda et al, 1995) as a temporary covering graft (instead of cadaveric allogenic skin) for managing extensive burns before permanent autografting (Allen et al, 1994; Davis et al, 1995; Hansbrough, 1995; Hansbrough et al, 1994; Ilten-Kirby et al, 1995), and as a therapeutic acid for the healing of chronic skin ulcers (Black et al, 1994; Gentzkow et al, 1994).

The basis of Dermagraft products is the three-dimensional cultivation human diploid fibroblast cells, which are grown on a polymer scaffold (Figure 1) and which secrete a mixture of growth factors and matrix protein (Figures 2 and 3) (Landeen et al, 1992). To produce Dermagraft, human fibroblast cell strains are established from surgically removed neonatal foreskins and cultured by standard methods (Jacoby and Pastan, 1979; Kruse and Patterson, 1973). Maternal blood samples are tested for exposure to infectious diseases, including human immunodeficiency virus (HIV), human T-cell lymphotropic virus (HTLV), herpes simplex virus (HSV), cytomegalovirus, and hepatitis. An initial screen is made of the cultured cells for sterility, mycoplasma, and for eight human viruses: adeno-associated virus, HSV 1 and 2, CMV, HIV 1 and 2, and HTLV I and II. Master cell banks (MCB) are created at Passage 3, and manufacturer's working cell banks (MWCB) are created at Passage 5. The cell banks undergo extensive additional testing (Table 1) according to applicable sections of the FDA "Points to Consider" (U.S.F.D.A., 1993) and the guidelines from the European Union Committee for Proprietary Medicinal Products (CPMP) (CPMP, 1989).

Dermagraft products are produced by seeding the polymer scaffolds at Passage 8 at a population doubling level of approximately 30 (which is at

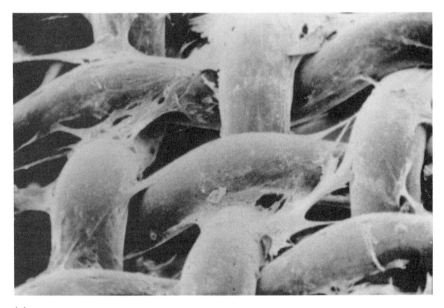

(a)

(b)

Figure 1. (a) Scanning electron micrograph (SEM) of fibroblasts stretching across the mesh substrate 1 to 2 days after seeding. (b) Photomicrograph (40X) of confluent cells on the mesh substrate 4 to 5 days after seeding.

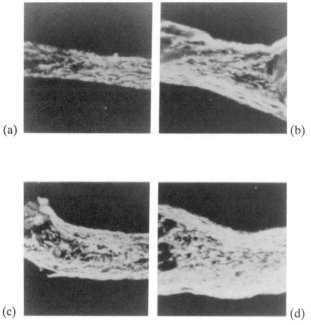

(a)   (b)

(c)   (d)

Figure 2. Photomicrographs (400X) showing increase in fibronectin immunofluorescence of dermal replacement tissue during culture: (a) 1 week, (b) 2 weeks, (c) 3 weeks, and (d) 4 weeks.

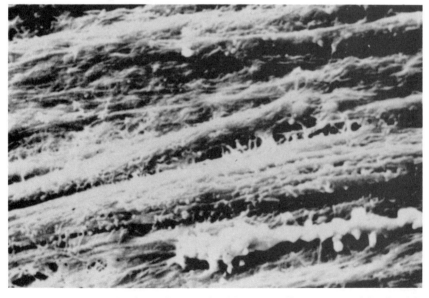

Figure 3. Scanning election micrograph of human collagen secreted in the 3-D culture system.

Table 1. Tests applied to human tissues used for establishment of master cells banks for tissue engineering.

Test	Prime Concern
In Vitro Tumorigenicity	Cancer potential
Karyology	cells course, cancer potential
Isoenzyme Analysis	cell source
Fluorescent Antibody Staining	cell source
DNA Profiling	cell source, reference for cell identity
Sterility Testing for Bacterial and Fungal Contaminants	Sterility
Mycoplasma	Sterility
In Vivo Assay for Adventitious Viral Contaminants	viral contamination
Hepatitis B & C via PCR	viral contamination
Cytomegalovirus (CMV) via PCR and Co-Cultivation	viral contamination
In Vitro Assay for Adventitious Viral Contaminants	viral contamination
Electron Microscopy	viral contamination
Dual Template Reverse Transcriptase Assay	retroviral contamination
HIV 1 & 2 via PCR and Co-Cultivation	retroviral contamination
Human T-cell Lymphotropic Virus Type II by PCR	retroviral contamination

about half the life-span for this cell type). With this procedure, a single foreskin donor provides sufficient cell seed to produce a 250,000 ft^2 of finished Dermagraft product. This accepted procedure for producing biologic products derived from human cells for medical application (Petricciani and Hennessen, 1987; Petricciani et al, 1979) provides a much higher level of safety, uniformity, and availability than is possible with the minimally controlled and tested cadaveric tissues from multiple donors currently used for implantation (Bale et al, 1992; Blood et al, 1979; White et al, 1991).

The cells are seeded initially at a fairly high density of $1–3.10^5/cm^2$ of superficial area. During the following 4–7 days the population increases rapidly in a sigmoidal fashion reaching a final density of $0.8–1.5.10^6$ cells/cm^2 (Figure 4). By contrast, little collagen is deposited during the first week, but from day 7 onwards, matrix is produced at a rate that increases until the time of harvest at 16–25 days. Presumably the rate of collagen deposition eventually levels out, but Dermagraft is normally harvested while its rate is very high. Thus, the accumulation of culture components can be broadly divided into two categories: accumulation like cells, rapid and sigmoidal, and accumulation like collagen, with a long lag.

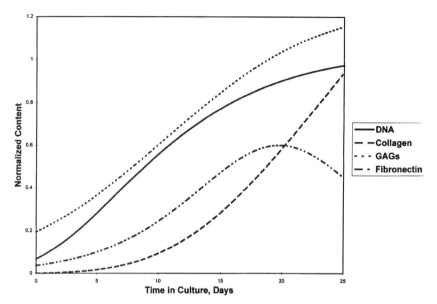

Figure 4. Deposition of tissue constituents in Dermagraft three-dimensional human fibroblast culture.

Human foreskin fibroblasts, expanded in monolayer culture for 8 passages, were seeded onto Biobrane in continuous-flow bioreactors at a density of about 200,000 cells/cm^2. Bioreactors were fed with DMEM containing 10% calf serum at a rate of 1.5 cm/min, with a residence time of 11 min and a total recirculation volume of 250 mL/bioreactor. Medium was changed on days 7, 10, 13, 16, 19, and 22. The DNA content of the tissue was estimated by Hoechst 33258 binding, collagen by Sirius Red binding, glycosaminoglycan by dimethylmethylene blue metachromasia, and fibronectin by fluorescent ELISA. The results represent the consolidated results of 6 independent time courses. The lines were derived by fitting the 11–50 data points for each component to the Gompertz equation by the method of least squares.

Other tissue components have been examined in regard to their manner of deposition, notably glycosaminoglycans and fibronectin. Glycosaminoglycans comprise a heterogeneous group of molecules that carry repeating disaccharide chains attached to serine or threonine residues. The carbohydrate side chains are frequently sulfated and are important in growth factor action. A prominent member of this group of molecules in Dermagraft is decorin, which binds to collagen and is incorporated into the extracellular matrix. Others include versican, an extracellular matrix protein, and syndecan and betaglycan which are cell membrane proteins. Glycosaminoglycans thus occur both as extracellular matrix and as cellular components. Their pattern of synthesis, however, is clearly related to the cellular category. Sections stained for glycosaminoglycans with Alcian blue show a substantial carbohydrate extracellular matrix before appreciable collagen has been deposited.

Fibronectin is an important cell adhesion molecule in wound healing (French et al, 1989). In addition to allowing cell attachment, spreading and migration, it also transmits growth factor-like information to the cellular signal transduction system (Masur et al, 1995), largely through the $\alpha_5\beta_1$ integrin (the fibronectin receptor). Fibronectin mRNA is spliced to generate a range of variants that differ in expression in a tissue specific manner during developmental and during wound healing (Schwarzbauer, 1991). During Dermagraft culture, fibronectin containing both variably spliced fibronectin type III repeats (the IIIB and IIIA) regions is formed, as in healing wounds, but all five possible splice variants of the V region, some of which include the $\alpha_4\beta_1$ integrin binding sequence, are produced. This integrin occurs on circulating cells of the immune system, and the controlled expression of its ligand is potentially of importance in modifying diapedesis. In terms of its synthesis, fibronectin falls into the matrix category, being made at a high rate after a lag. Unlike other matrix proteins, the fibronectin content of Dermagraft reaches a maximum at the time at which Dermagraft is harvested (16–21 days) and shows evidence for a declining concentration at later times.

Cellular proliferation and matrix deposition in the three-dimensional culture system is presumably controlled by growth factors. Known growth factors that influence these processes are PDGF and TGFβ. The PDGF system in culture operates in an autocrine fashion. The fibroblasts express the short form of PDGF A chain that does not interact with the extracellular matrix and is freely diffusible (Kelly et al, 1993; Pollock and Richardson, 1992). The messenger RNA is highly expressed in monolayer cultures but rapidly reduced (within 1 day) to a lower level on seeding in three-dimensional cultures (Figure 5). The cells can be stained immunohistochemically for the PDGF α receptors, which is activated by PDGF A chain, and also for the PDGF β receptor (Figure 6). Since the receptors are detectable at all times during the culture, we infer that they are not down-regulated and that the autocrine system is operating in ligand-limited mode. Thus the cells are potentially sensitive to the concentration of the autocrine ligand which, since it is freely diffusible, will be influenced by cell density and the configuration of the space surrounding them. Examination of PDGF AB-induced tyrosine autophosphorylation of the PDGF receptor has shown that the cells are only responsive during the first few days of culture, when proliferation is very high. At times later than 7 days, PDGF receptor tyrosine phosphorylation is not detectable after PDGF AB addition, even in the presence of vanadate. Our working hypothesis is that the PDGF autocrine system provides information to the cells on the local cellular environment during the proliferative phase of the culture.

Collagen synthesis is known to be stimulated by TGFβ in three-dimensional fibroblast culture. In the Dermagraft system, we have found mRNA

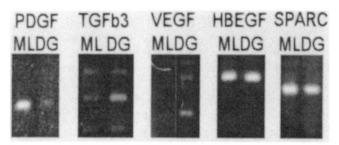

Figure 5. Comparison of the expression of growth factors in monolayer and three-dimensional Dermagraft culture. Culture was performed as described under Figure 1. RNA was extracted by an SDS precipitation procedure (Amresco, Solon, OH). The RNA was reverse transcribed using Superscript Reverse Transcriptase (Life Technologies, Gaitersburg, MD) and random hexamer primers. The same batch of reverse transcript was used for detection of all the growth factors. PCR was performed under standard conditions, using 4 µL reverse transcript, corresponding to 200 ng RNA in a total volume of 20 µL. PCR conditions of PDGF A chain were 55°C annealing, 29 cycles; TGFβ3, 50°C annealing, 40 cycles; VEGF, HBEGF and SPARC, 55°C annealing, 35 cycles. ML, monolayer; DG, Dermagraft.

expression for $TGF\beta_1$ to be at the limit of detection of the RT-PCR system used. However, the protein is detectable in medium and tissue by immunoblotting. $TGF\beta_1$ is secreted in a latent form that is deposited in the matrix through a specific binding protein. Activation involves thrombospondin or plasmin and the IGF-2 receptor (Nunes et al, 1995). Since we can detect accumulation of the protein, we infer that it remains as the latent form and is not active in controlling cellular activity during culture. By contrast, $TGF\beta_3$ mRNA is induced severalfold on seeding fibroblasts in three-dimensional culture (Figure 5), although we have been unable to detect the protein by immunoblotting. Our working hypothesis is that it is activated and consumed by interaction with receptor or other mechanisms and that it plays a role in the delayed induction of collagen synthesis.

In addition to the growth factors thought to act in an autocrine manner during the culture process, expression of several paracrine factors has been detected. These include both angiogenic factors, such as VEGF and SPARC, and factors expected to activate keratinocytes such as KGF and HBEGF.

The VEGF gene gives rise to four splicing variant mRNAs which are translated to VEGF molecules of different lengths (Houck et al, 1992). The regions deleted in the smaller molecules carry heparin binding sites so that the longer members of the group bind progressively more tightly to extracellular matrix while the shorter are freely diffusible. Although all four variants can be detected in Dermagraft RNA preparations, the predominant ones are those corresponding to the 121 amino acids and 189 amino acids molecules (Figure 5). We expect the first of these to be washed from the tissue, but the

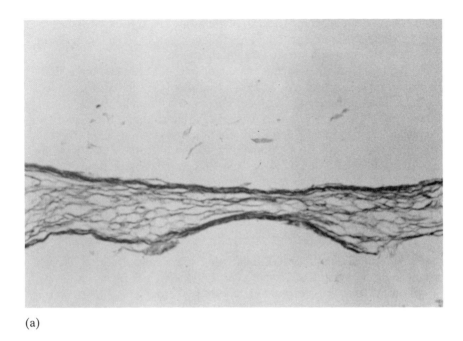

(a)

(b)

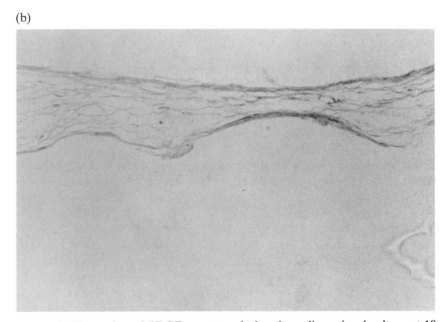

Figure 6. Expression of PDGF receptors during three-dimensional culture at 18 days. A, α receptors; B, β receptors.

second should remain bound to extracellular matrix. Basic FGF, which is a highly angiogenic, matrix binding growth factor can be detected in Dermagraft tissue by immunohistochemical methods.

SPARC is an antiadhesive and antiangiogenic molecule that binds PDGF BB and albumin (Lane and Sage, 1994). Its proteolytic breakdown products, however, include the highly angiogenic, copper-binding peptide gly-his-lys (Lane et al, 1994). SPARC mRNA is readily detected at all stages of Dermagraft culture (Figure 5), and a protein of molecular weight 43,000 reacts with anti-SPARC antibody on immunoblots. The role of SPARC in Dermagraft placed in wound sites is likely to depend on the proteolytic and inflammatory events. In the absence of such processes, SPARC is antiangiogenic and inhibits the activity of PDGF B chain, derived from platelets, while having little effect on the autocrine fibroblast PDGF A chain. If hydrolyzed in an inflammatory context, it will generate the angiogenic gly-his-lys peptide.

KGF is a member of the FGF family that stimulates keratinocyte proliferation and migration (Aaronson et al, 1991), which is highly induced in fibroblasts surrounding an incisional wound (Werner et al, 1992), KGF is expressed by fibroblasts in monolayer and three-dimensional culture, although the amount in three-dimensional culture is substantially reduced within three days of seeding. HBEGF mRNA is readily detected throughout the culture of Dermagraft (Figure 5) and in fresh, harvested product by RT-PCR. Both these growth factors bind to extracellular matrix, and we expect them to be present in the therapeutic product.

DERMAL REPLACEMENT FOR SEVERELY BURNED PATIENTS

Problems with Current Techniques for Treating Severely Burned Patients

Currently, when a patient is so severely burned that there is not enough unburned skin for an autograft, surgeons use human cadaveric skin, either fresh or cryopreserved, as a biologic dressing for the excised wound bed (Hansbrough, 1987; 1995; Hansbrough et al, 1993; 1994). Additional autograft skin becomes available by repeated cycles of harvest and regrowth from unburned areas of the patient; the allograft is replaced as soon as possible by this additional autograft. However, by its nature, this procedure carries with it several problems. First, the cadaveric allograft consists of both dermal and epidermal skin layers and is generally antigenic, which results in graft rejection within a few days or a few weeks. The graft rejection necessitates additional surgery to replace the failed allograft. Moreover, this process must be repeated until sufficient autograft skin is available to completely cover the burned area. Second, the availability of cadaveric allograft skin is limited. Third, the potential of cadaveric allograft skin to transmit disease has been recognized as an increasing problem since late 1993, when the United States Food and Drug Administration issued an interim rule (21 CFR Part 1270) that detailed the requirements for "banked human tissue" to prevent the transmission of acquired immune deficiency syndrome (AIDS) and hepatitis through human tissue used in transplantation (US FDA, 1993b).

Bioengineering a Superior Substitute for Cadaver Allograft

The tissue engineered product, Dermagraft-Transitional Covering (Dermagraft-TC), has been developed to provide a disease-free, longer lasting alternative to cadaveric autografts in severely burned patients. Human fibroblast cells prepared as described previously are seeded onto a nylon mesh attached to a thin silicon rubber membrane (Biobrane™, Dow B. Hickam, Sugarland, TX). The nylon forms the three-dimensional scaffolding for growth of the Dermagraft tissue, and the silicon membrane forms an artificial (nonimmunogenic) epidermis to prevent fluid loss (Figure 7). As the cells grow in culture, they secrete proteins and growth factors in a balanced mixture that is natural for these types of cells, thereby generating a three-dimensional tissue matrix (Mansbridge et al, 1995). At the conclusion of the growth period, the closed production system is separated so that the individual bioreactors or cassettes used to grow the product are sealed and thus become the product's primary package. The sealed cassettes are frozen and stored at low temperatures (below $-70°C$). The bioreactor/final

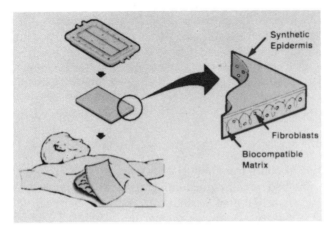

Figure 7. Drawing of Dermagraft-TC as it is applied to the excised wound bed to stimulate neovascularization.

Figure 8. Photograph of Dermagraft-TC as it is being removed from the bioreactor cassette where it is grown, frozen, stored, and shipped to physicians.

product package system ensures delivery to the physician in an easily accessible form, providing the highest level of safety (Figure 8). Sterility assays ensure the product's safety; assays are run on final-production medium aliquots that have been exposed to every product unit, and destructive final-product-release testing is performed on a statistically appropriate number of units per lot.

In the clinical pilot study of Dermagraft-TC, a total of ten patients were enrolled at three burn centers. For each patient, one site was treated with Dermagraft-TC and a similar control site was treated with cryopreserved cadaveric allograft (Davis et al, 1995). Preliminary clinical experience suggests that Dermagraft-TC provides a temporary wound cover as effective as cryopreserved allograft and that it may remain in place for as long as approximately 6 weeks (cadaveric allograft is generally rejected earlier, within 2–3 weeks). A pivotal clinical trial has been completed to obtain further supportive data regarding the safety and efficacy of Dermagraft-TC.

Mechanism

Current opinion in burn treatment is that optimal results are obtained if the debrided wound bed remains quiet until grafting with minimal development of granulation tissue. Human wounds covered with Dermagraft-TC show minimal inflammatory activity, as though the wound healing system has been shut down. Possible mechanisms contributing to this include the range of growth and matricellular factors present. Since the tissue is nonviable when applied and has been extensively washed, diffusible factors, such as PDGF A chain and the 121 residue form of VEGF, are not present. All factors in the tissue are bound and only available to invading cells. Such molecules include TGFβ$_1$, the 189 residue form of VEGF, and basic FGF. In the case of TGFβ$_1$, the latent form also has to be activated. In addition, several molecules are present that inhibit tissue activation such as SPARC, decorin, and tenascin. Intact SPARC is an antagonist of cell adhesion to fibronectin and is also able to moderate the effects of PDGF B chain and other growth factors. Decorin binds growth factors and has been reported to inhibit the activity of TGFβ$_1$ (Yamaguchi et al, 1990). Il-1α is not present at detectable levels in Dermagraft-TC as harvested, although its induction in fresh (viable) tissue by agents such as phorbol esters can be demonstrated. Similarly, although stromelysin (MMP-3) can be induced in fresh Dermagraft by IL-1α, this cannot occur in the nonviable product applied to patients. Overall, the presence of proteins that moderate growth factor activity in a selective manner, together with the absence of diffusible activating agents provide, a noninflammatory environment that remains responsive to outside factors.

Epithelialization and Tissue Engineered Dermal Replacements

Use of meshed, expanded, split-thickness skin grafts to cover full-thickness wounds can result in the full epithelialization of the mesh interstices (Economou et al, 1995; Hansbrough et al, 1992a). However, poor cosmetic and functional outcomes frequently occur even with full epithelialization because the epithelium that grows across the graft interstices lacks a dermis. When meshed, expanded, split-thickness skin grafts have been placed *over* Dermagraft, the tissue engineered dermal replacement, the resulting epithelialized layer covers a densely cellular substratum that is very similar to dermis (Hansbrough et al, 1992a). This result indicates that the combination of meshed, expanded, split-thickness skin grafts and Dermagraft may improve cosmetic and functional outcomes in patients who need allografts.

TREATMENT FOR SKIN ULCERS

The Special Problems and Importance of Treating Chronic Skin Ulcers

Preliminary studies, discussed below, indicate that the treatment of chronic skin ulcers, a severe medical problem, may be improved by tissue engineering technology. There are more than 2 million cases of chronic, slow healing or nonhealing skin ulcers treated in the United States each year. In these wounds, the skin breaks down as a result of disruption of blood flow caused either by prolonged pressure over a localized area or by chronic diseases that affect the circulatory or peripheral nervous systems. More than 500,000 patients are treated annually for chronic ulcers associated with diabetes, but healing is variable and the recurrence rate of ulcers at the same site is high. Costs for treating these chronically open wounds ranges from \$9,000 to \$60,000 per year (MedPRO, 1992), and current treatments are often inadequate, resulting in approximately 50,000 amputations per year among diabetic patients (CDC, 1991). Approximately 50% of all amputees die within 3 years of surgery (Bild et al, 1989).

Use of Tissue Engineered Dermal Replacement in Treating Chronic Skin Ulcers

A neonatal human fibroblast product, Dermagraft-Ulcer, which is similar in character to Dermagraft-TC, may improve dramatically the current therapeutic approaches to chronic skin ulcers. Fibroblast cells are inoculated onto a dissolvable polymer mesh (Vicryl™; Ethicon, Sommerville, NJ) for scaffolding; the cells are then allowed to grow into the three-dimensional tissue. The ultimate goal is to mimic growth conditions that allow the formation of a normal dermis, thus simulating normal dermal growth and avoiding conditions which would trigger a wound response which would result in dermal scars. This product is also frozen and can be stored long term until it is needed by physicians.

Figure 9(a) shows the application of this product to a diabetic foot ulcer. In a 50-patient pilot clinical trial, patients receiving the eight-dose regimen showed statistically significant improvement ($P < .05$); complete wound closure was seen in 50% of the Dermagraft patients and in only 8% of the patients in the control group (who received conventional therapy) (Figure 9b). In the patients with Dermagraft-treated ulcers that healed, no ulcer recurrence was seen during follow-up ranging from 4 months to 1 year. Pivotal clinical trials are now under way in both the United States and France to further substantiate these results. In a separate clinical trial of patients with venous ulcers, recurrence of healed ulcers was 6.3% for Dermagraft-treated patients versus 19.7% for control patients ($P < .05$). Dermagraft-Ulcer has also been shown to be effective in treating pressure ulcers, the most common type of these chronic wounds. In a pilot trial of 50 patients with pressure ulcers, complete healing was achieved in 46% of patients receiving the eight-dose Dermagraft regimen and in only 25% of control patients; wound closure was evaluated over a period of 12 weeks after the treatment period. Interim data from a blind controlled pivotal trial

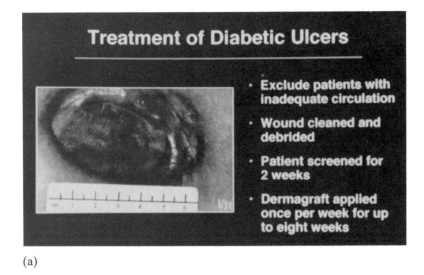

(a)

(b)

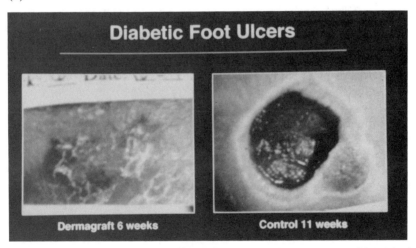

Figure 9. (a) Application of Dermagraft to a debrided ulcer wound. (b) Diabetic ulcer treated with Dermagraft showing complete epithelialization in contrast to unhealed control ulcer.

for the treatment of diabetic ulcers has been encouraging and the pivotal trial is on track for near-term completion.

Mechanism

By contrast with the treatment of burn wounds, the goal of ulcer treatment is to stimulate a wound healing response. Comparison of the conditions under which Dermagraft is used in the two situations provides some insight into possible mechanisms of action, although we are far from understanding the complexities of wound healing. Dermagraft-U is viable as applied, and its fibroblasts persist in the wound site. Thus the diffusible products of the cells, such as PDGF A chain and the 121 residue VEGF, continue to be produced and spread into the wound bed, potentially stimulating cell proliferation, chemotaxis, and angiogenesis. The fibronectin in Dermagraft is of the type that is found in wounds and supports cell migration, including a proportion of the splicing variants that reacts with the $\alpha_4\beta_1$ integrin found on circulating cells. Fibroblasts around the ulcer express α and β PDGF receptors (Peus et al, 1995) and thus possess receptors confering the potential to respond to the PDGF A chain made by the Dermagraft cells. At the same time, both wound and Dermagraft cells have receptors for PDGF B chain derived from platelets. As cells move into Dermagraft, bound growth factors, such as HBEGF, basic FGF, and KGF become available. Ulcer beds contain high concentrations of metalloproteases (Falanga et al, 1994; Wysocki et al, 1993), including stromelysin that will degrade SPARC. Degradation products of SPARC include angiogenic peptides based on gly-his-lys that have been shown to aid in the healing of diabetic ulcers (Mulder et al, 1994). The matrix of Dermagraft-U contains $TGF\beta_1$ in latent form that can be activated, and the Dermagraft fibroblasts express $TGF\beta_3$.

In addition to containing or being able to produce a series of cytokines controlling cell migration and proliferation, matrix deposition, angiogenesis, and reepithelialization, Dermagraft cells also contain the systems for the control of their production. Over-stimulation of many receptors causes downregulation and consequent insensitivity to ligand. It was noted in the discussion of PDGF during Dermagraft culture in which the concentration of PDGF A chain is largely determined by the cells that activation of the autocrine system is ligand limited and the receptors remain expressed. In support of the same point of view, experiments on the effects of TGFb1 on wound healing in rabbit ears have shown that low doses are effective in increasing matrix deposition, but that at high doses the result is reversed (Beck et al, 1989). In addition to the ability to maintain appropriate growth factor concentrations, the cells are also able to respond to stimuli originating from external events such as inflammation or platelet lysis.

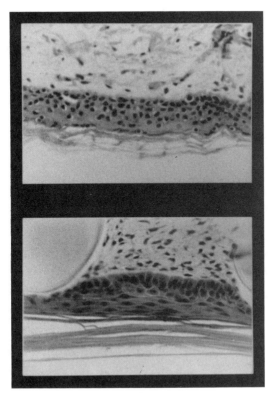

Figure 10. Cross-sectional comparison of human skin and Skin²™ ZK1300 model: hematoxylineosin staining.

Overall, Dermagraft-U brings to the ulcer site fresh fibroblasts, unmodified by exposure to the ulcer environment, and some bound growth factors, together with the ability to produce diffusible factors continuously at concentrations that lie within the physiological range and under the control of cellular delivery mechanisms.

IN VITRO TOXICOLOGY TESTING

An estimated 10,000 new chemical entities are produced each year; the manufacturers of these new chemicals need effective and economical safety testing procedures. At the same time as the need for testing is increasing, societal and regulatory pressures are compelling chemical and cosmetic producers to reduce or replace the use of animals as a predictive model for assessing human toxicity to environmental factors (Mayer et al, 1994). This search for effective, economical, nonanimal testing methods for new compounds has encouraged the development of reliable and well-standardized in vitro methods for toxicological testing.

Because of the complexity of the human skin and its functions, development of an accurate nonanimal *in vitro* alternative model has been challenging. The human skin serves as a barrier protecting the internal environment against physical damage and against external factors such as bacteria, viruses, environmental toxins, and ultraviolet light. For the epidermis to produce the most effective toxicological, immunological, and biochemical barrier, the major cell types of the epidermis (keratinocytes) and the dermis (fibroblasts) must function together in a dynamic, integrated fashion (Rheins et al, 1994; Zeigler et al, 1993). The epidermal-dermal intercellular biochemical signals such as interleukins, cytokines and other factors provide the skin with local homeostatic signals to ensure the integrity of that skin in response to a variety of environmental insults.

Using its proprietary technology, Advanced Tissue Sciences has successfully developed a three-dimensional human skin tissue model called skin[2]®. The *in vitro* model is tissue engineered to mimic the skin both biochemically and morphologically (Figure 10). This skin tissue system is composed entirely of metabolically and mitotically active dermal and epidermal layers supported by a scaffold of nylon mesh. These layers are produced in culture by initial seeding of the nylon mesh with neonatal foreskin fibroblasts that secrete functionally active extracellular matrix proteins, glycosaminogylycans, and growth factors (Zeigler et al, 1993). This first seeding process results in a dense dermal tissue which is then seeded with neonatal foreskin keratinocytes. The keratinocytes rapidly attach to the dermal tissue, proliferate and differentiate into a functional multilayered epidermis including stratum corneum (DeWever and Rheins, 1994; Fleischmajer et al, 1991). This skin model has been used to study the development of various supramolecular structures including elastin-associated microfibrils and collagen fibrillogenesis (Zeigler et al, 1993). The system allows the cells to develop an architecture and metabolic capacity at the molecular level that resembles the in vivo tissue (Fleischmajer et al, 1991; 1993; Rheins et al, 1994; Zeigler et al, 1993).

These structural and biochemical similarities to normal human tissue explain the success of Skin[2] in a wide variety of in vitro testing protocols (Beck et al, 1989; Casterton et al, 1994; DeWever and Rheins, 1994; Edwards et al, 1994; Koschier et al, 1994). The small pieces of tissue provided can be challenged by topical applications of substances whose irritant potential is measured by the time course of cellular enzyme release production of inflammatory mediators and by residual cell viability. These materials have been used commercially by many companies for screening the safety of chemical ingredients and formulations in development of their finished products before release to the public.

Validation studies for predicting human safety in blind, side-by-side comparisons with animal methods such as the Draize ocular irritation assay have been conducted in the United States and Europe. An example of such a formal validation process is the 3-year study being conducted by the European Commission and the European Cosmetic Association on in vitro alternatives including Skin² to evaluate the phototoxic potential of personal and skin care products. In this study, various concentrations of test chemicals are applied to the epidermal side of the test tissue for contact times of 1 or 24 hours followed by exposure to 2.9 J/cm² of ultraviolet-A radiation. After exposure, the test substrates are rinsed free of test chemicals and allowed to recover for 24 hours. Cytotoxicity is measured by a mitochondrial assay to determine which chemicals can mask the phototoxic reaction. Current results have shown that data from Skin² studies have a high degree of correlation with data from human and animal models now used to evaluate the phototoxic potential of chemicals (Edwards et al, 1994).

The United States Department of Transportation and Transport Canada, have approved the use of Skin² as an in vitro laboratory test alternative to live animal testing for potentially corrosive materials. The testing requirement was mandated by a United Nations directive for transportation of hazardous materials. This recognition is an important milestone; it is the first United States government regulatory approval for a human tissue engineered product. Extensive, worldwide studies are proceeding to obtain further regulatory acceptance on these alternative testing approaches. The most recent of these is a validation study comparing the release of the inflammatory mediator, interleukin 1 alpha from skin²® with the 14 day cumulative irritation in human volunteers following application of surfactants.

In addition to these product safety tests utilizing skin²®, a series of protocols have been developed to measure the efficacy of topically applied materials on cutaneous physiology (Slivka, 1992). These applications take advantage of and rely upon the cell cell interactions that occur between the keratinocytes and fibroblasts which constitute the epidermal and dermal layers, respectively (Slivka et al, 1992). Examples include the regulation of extracellular matrix constituents, (Figure 11) in response to pharmaceuticals or cosmetic formulations such as retinoids and the alphahydroxy acids (Donnelly et al, 1995). Numerous companies are utilizing this approach with skin²® to evaluate and compare ingredients and final formulations for efficacy and product claims. These tests offer an economical alternative to human clinical trials in the early stages of product development.

CONCLUSION

As a rapidly emerging technology, tissue engineering holds the promise of an entirely new approach to the repair and reconstruction of tissues and

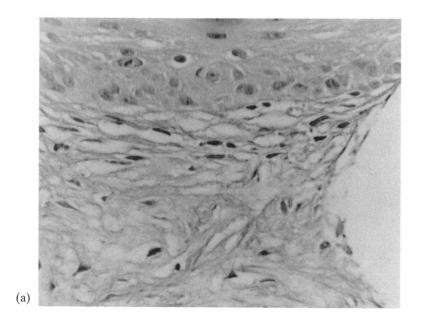

(a)

(b)

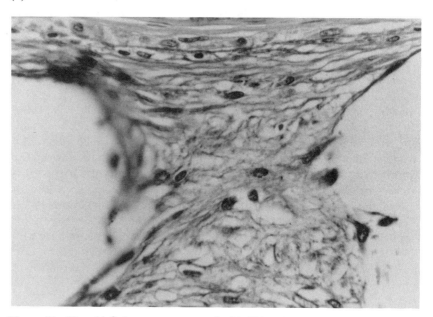

Figure 11. The skin² tissues were treated with 100 mg/ml sodium ascorbate in the medium for 3 days with and without topical application of an alpha hydroxy acid formulation. The tissues were then fixed and stained for Type I Collagen by immunohistochemistry. Note the increase in the staining in the tissue treated with the alpha hydroxy acid formulation denoting an increase in Collagen Type I deposition.

a: Sodium ascorbate control
b: Alpha hydroxy acid formulation

organs damaged by disease, accident, or congenital abnormalities and defects. The advantages of using normal human tissues produced by in vitro culture techniques include the safety, efficacy of the product and availability of a consistent material. Tissue engineering holds the promise of combining the advances in culture technology with progress in medical and surgical intervention to provide new solutions by implantation of normal human tissue constructs that are best able to perform the natural functions for each tissue type.

The therapeutic applications reported here—the use of dermal fibroblast-based products for management of accidental or disease-related skin damage—demonstrate that medical intervention by implantation of in vitro engineered human tissue is a feasible and effective treatment. In ongoing clinical studies, tissue engineered skin replacements are currently being tested for treatments of patients who are severely burned or who have chronic skin ulcers. Results so far indicate that these dermal replacements have great potential to improve the therapeutic choices for these patients.

To increase the use of tissue engineered skin for toxicology testing and thereby decrease the use of animals in such testing, extensive regulatory work is ongoing. These products have the potential to significantly help protect the individual from the increasing array of new chemicals entering his environment by providing a more economical and effective method to pre-screen for harmful properties of the new materials.

Tissue engineered dermis, therefore, may have a dramatic effect not only on the therapies available to injured patients but also on the effectiveness of toxicological testing to prevent skin damage from chemical and environment damage.

References

Aaronson SA, Bottaro DP, Miki T, et al (1991): Keratinocyte growth factor. A fibroblast growth factor family member with unusual target cell specificity. *Ann NY Acad Sci* 638:62–77

Allen LA, Landeen LK, Pieters R, et al (1994): Development of a porcine full-thickness burn model for the evaluation of human cadaveric skin and Dermagraft Transitional Covering, a biosynthetic skin substrate. In: *Proceedings of the 9th Congress of the International Society for Burns*. Paris, France

Bale JF, Kealey GP, Ebelhack CL, et al (1992): Cytomegalovirus infection in a cyclosporine-treated burn patient: A case report. *J Trauma* 32:263–267

Beck LS, Chen TL, Hirabayashi SE, et al (1989): Accelerated healing of ulcer wounds in the rabbit ear by recombinant human transforming growth factor-β1. *Growth Factors* 2:273–282

Bell E, Ivarsson B, Merril C (1979): Production of a tissue-like structure by construction of collagen lattices by human fibroblasts of different proliferative potential in vitro. *Proc Natl Acad Sci USA* 76:1274–1278

Bild DE, Selby JV, Sinnock P, et al (1989): Lower-extremity amputation in people with diabetes: Epidemiology and prevention. *Diabetes Care* 12:24

Black K, Gentzkow G, Purchio T, et al (1994): The future for temporary and permanent wound coverage. In: *Proceedings of the 3rd International Congress on the Immune Consequences of Trauma, Shock and Sepsis.* Munich, Germany

Blood S, Heck E, Baxter CR (1979): The importance of the bacterial flora in cadaver homograft burn donor skin. In: *Proceedings of the 11th Meeting of the American Burn Association.* New Orleans, LA

Casterton PL, Potts LF, Klein BD (1994): Use of in vitro methods to rank surfactants for irritation potential in support of new product development. *Toxicol In Vitro* 8:835–836

Centers for Disease Control (1991): *The Prevention and Treatment of Complications of Diabetes: A Guide for Primary Care Practitioners,* NIH Publication No. 93-3464. Atlanta, GA: Centers for Disease Control

Committee for Proprietary Medicinal Products: Ad Hoc Working Party on Biotechnology/Pharmacy (1989): Notes to applicants for marketing authorizations on the production and quality control of monoclonal antibodies of murine origin intended for use in man. *J Biol Stand* 17:213

Contard P, Bartel RL, Jacobs L, et al (1993): Culturing keratinocytes and fibroblasts in a three-dimensional mesh results in epidermal differentiation and formation of a basal lamina-anchoring zone. *J Invest Dermatol* 100:35–39

Cuono C, Langdon R, McGuire J (1986): Use of cultured epidermal autografts and dermal allografts as skin replacement after burn injury. *Lancet* 1(8490):1123–1124

Cuono CB, Langdon R, Birchall N (1987): Composite autologous-allogenic skin replacement: Development and clinical application. *Plast Reconstr Surg* 80:626

Davis MM, Hansbrough JF, Mozingo DW, et al (1995): Conference Center, Harrogate Dermagraft-TC in the treatment of burns requiring temporary coverings. In: *Proceedings of the Symposium on Advanced Wound Care and Medical Research Forum on Wound Repair.*

Demarchez M, Hartmann DJ, Regnier M, et al (1992): The role of fibroblasts in dermal vascularization and remodeling of reconstructed human skin after transplantation onto the nude mouse. *Transplantation* 54:317–356

DeWever B, Rheins LA (1994): Skin2™: An in vitro human skin analog. In: *Alternative Methods in Toxicology, Vol. 10,* Rougier A, Goldberg AM, Maibach HI, eds. New York: Mary Ann Leibert, Inc

Donnelly TA, Allen R, Edwards S, Rheins LA (1995): Dermal Matrix Deposition in an in vitro skin model following topical administration of alpha hydroxy acid (AHA) formulations. *J Invest Derm* 104:669

Economou TP, Rosenquist MD, Lewis II RW, et al (1995): An experimental study to determine the effects of Dermagraft on skin graft viability in the presence of bacterial wound contamination. *J Burn Care Rehab* 16:27–30

Edwards SM, Donnelly TA, Sayre RM, et al (1994): Quantitative in vitro assessment of phototoxicity using a human skin model, Skin2™. *Photodermatol Photoimmunol Photomed* 10:111–117

Falanga V, Grinnell F, Gilchrest B, et al (1994): Workshop on the pathogenesis of chronic wounds. *J Invest Dermatol* 102:125–27

Fleischmajer R, Contard P, Schwarz E, et al (1991): Elastin-associated microfibrils (10 nm) in a three-dimensional fibroblast culture. *J Invest Dermatol* 97:638–643

Fleischmajer R, MacDonald ED, Contard P, et al (1993): Immunochemistry of a keratinocyte-fibroblast co-culture model for reconstruction of human skin. *J Histochem Cytochem* 41:1359–1366

French CC, Van de Waters L, Dvorak H, et al (1989): Reappearance of an embryonic pattern of fibronectin splicing during wound healing in the adult rat. *J Cell Biol* 109:903–914

Gentzkow G, Iwasaki S, Gupta S, et al (1994): *Cultured Human Dermal Replacement Tissue for the Treatment of Diabetic Foot Ulcers.* Oxford, England: European Tissue Repair Society

Hansbrough JF (1987): Biological dressings. In: *The Art and Science of Burn Care,* Boswick J, ed. Rockville, MD: Aspen Publishers

Hansbrough JF (1990): Current status of skin replacements for coverage of extensive burn wounds. *J Trauma* 30:S155–162

Hansbrough JF (1995): Status of cultured skin replacements. *Wounds* 7:130–136

Hansbrough JF, Cooper ML, Cohen R, et al (1992a): Evaluation of a biodegradable matrix containing cultured human fibroblasts as a dermal replacement beneath meshed skin grafts on athymic mice. *Surgery* 4:438–446

Hansbrough JF, Dore C, Hansbrough WC (1992b): Clinical trials of a living dermal tissue replacement placed beneath meshed, split-thickness skin graft on excised burn wounds. *J Burn Care Rehab* 13:519–529

Hansbrough JF, Morgan J, Greenleaf G (1993): Advances in wound coverage using cultured cell technology. *Wounds* 5:174–194

Hansbrough JF, Morgan J, Greenleaf B, et al (1994): Development of a temporary living skin replacement composed of human neonatal fibroblasts cultured in Biobrane, a synthetic dressing material. *Surgery* 115:633–644

Houck KA (1992): Dual regulation of vascular endothelial growth factor bioavailability by genetic and proteolytic mechanisms. *J Biol Chem* 267:26031–26037

Ilten-Kirby B, Pinney E, Pavalec R, et al (1995): Human allograft vs. Dermagraft Transitional Covering, a tissue engineered skin substitute for use as a full-thickness burn covering. *Proceedings of the 2nd Annual Conference on Cellular Engineering.* San Diego, CA: International Federation for Medical and Biological Engineering

Jakoby WB, Pastan IB, eds. (1979): Cell Culture. In: *Methods in Enzymology.* 58 New York: Academic Press

Kelly JL (1993): Accumulation of PDGF B and cell-binding forms of PDGF A in the extracellular matrix. *J Cell Biol* 121:1153–1163

Koschier FJ, Roth RN, Stephens TJ (1994): In vitro skin irritation testing of petroleum-based compounds in reconstituted human skin models. *J Toxicol-Cut Ocular Toxicol* 13:23–37

Kruse Jr PF, Patterson Jr MK, eds. (1973): *Tissue Culture Methods and Applications.* New York: Academic Press

Landeen LK, Zeigler FC, Halberstadt C, et al (1992): Characterization of a human dermal replacement. *Wounds* 5:167–175

Lane TF, Sage EH (1994): The biology of SPARC, a protein that modulates cell-matrix interactions. *FASEB J* 8:163–173

Lane TF, Iruela-Arispe ML, Johnson RS, et al (1994): SPARC is a source of copper-binding peptides that stimulate angiogenesis. *J Cell Biol* 125:929–943

Langer R, Vacanti JP (1993): Tissue engineering. *Science* 260:920–926

Leary T, Jones PL, Appleby M, et al (1992): Epidermal keratinocyte self-renewal is dependent upon dermal integrity. *J Invest Dermatol* 99:422–430

Lucas-Clerc C, Massart C, Campion JP, et al (1993): Long-term culture of human pancreatic islets in an extracellular matrix: Morphological and metabolic effects. *Mol Cell Endocrinol* 94:9–20

Mansbridge M, Gould T, Pinney E, et al (1995): Control of growth and matrix deposition in three-dimensional culture of human fibroblasts. In: *Proceedings of the 2nd Annual Conference on Cellular Engineering.* San Diego, CA: International Federation for Medical and Biological Engineering

Masur SK, Idris A, Michelson K, et al (1995): Integrin-dependent tyrosine phosphorylation in corneal fibroblasts. *Invest Ophthalmol Vis Sci* 36:1837–1846

Mayer FL, Whalen EA, Rheins LA (1994): A regulatory overview of alternatives to animal testing: United States, Europe, and Japan. *J Toxicol-Cut Ocular Toxicol* 13:3–22

MedPRO Monthly (1992): Wound healing: Focus on growth factors, skin substitutes. *MedPRO Month* 2:91

Mulder GD, Patt LM, Sanders L, et al (1994): Enhanced healing of ulcers in patients with diabetes by topical treatment with glycyl-L-histidyl-L-lysyl copper. *Wound Rep Reg* 2:259–269

Nunes I, Shapiro RL, Rifkin DB (1995): Characterization of latent TGF-beta activation by murine peritoneal macrophages. *J Immunol* 155:1450–1459

Perkins MA, Osborne R (1993): Development of an in vitro method for skin corrosion testing. *J Invest Dermatol* 100:535

Petricciani JC, Hennessen W, eds. (1987): Cells, products, safety: Background papers for the WHO study group on biologicals. In: *Developments in Biological Standardization.* 68 Karger S, ed. Basel Switzerland

Petricciani JC, Hopps HE, Chapple PJ, eds. (1979): *Cell Substrates.* New York: Plenum

Peus D, Jungtäubl H, Knaub S, et al (1995): Localization of platelet-derived growth factor receptor subunit expression in chronic venous leg ulcers. *Wound Rep Reg* 3:273–283

Pollock RA, Richardson WD (1992): The alternative-splice isoforms of the PDGF A-chain differ in their ability to associate with the extracellular matrix and to bind heparin in vitro. *Growth Factors* 7:267–277

Rheins LA, Edwards SM, Miao O, et al (1994): Skin2™: An in vitro model to assess cutaneous immunotoxicity. *Toxicol In Vitro* 8:1007–1014

Rheinwald JG (1989): Methods for clonal growth and serial cultivation of normal human epidermal keratinocytes and mesothelial cells. In: *Cell Growth and Division: A Practical Approach,* Baserga R, ed. New York: Oxford University Press

Schwarzbauer JE (1991): Alternative splicing of fibronectin: Three variants, three functions. *Bioessays* 13:527–533

Slivka SR (1992): Testosterone metabolism in an in vitro skin model. *Cell Biol Tox* 8:267–277

Slivka SR, Landeen LK, Zeigler F, Zimher MP, Bartel RL (1992): Characterization, barrier function and drug metabolism of an in vitro skin model. *J Invest Dermatol* 93:40–46

Takeda N, Abe S, Tsukiyama H, et al (1995): *The Clinical Experience of Artificial Skin Contained Allogenic Fibroblasts.* Nagoya, Japan: Japanese Society of Burn Injury

U.S. Food and Drug Administration (1993a): *Points to Consider in the Characterization of Cell Lines Used to Produce Biologicals.* Bethesda, MD: U.S. Department of Health and Human Services

U.S. Food and Drug Administration (1993b): *Inspectional Guidance: Banked Human Tissue.* Bethesda, MD: U.S. Department of Health and Human Services

Werner S, Peters KG, Longaker MT, et al (1992): Large induction of keratinocyte growth factor in the dermis during wound healing. *Proc Natl Acad Sci USA* 89:6896–6900

White MJ, Whalen JD, Gould JA, et al (1991): Procurement and transplantation of colonized cadaver skin. *Am Surg* 57:402–407

Wysocki AB, Staiano-Coico L, Grinnell F (1993): Wound fluid from chronic leg ulcers contains elevated levels of metalloproteinases MMP-2 and MMP-9. *J Invest Dermatol* 101:64–68

Yamaguchi Y, Mann D, Ruoslahti E (1990): Negative regulation of transforming growth factor-β by the proteoglycan decorin. *Nature* 346:281–284

Yannas IV, Burke JF, Orgill DP, Skrabut EM (1982): Wound tissue can utilize a polymeric template to synthesize a functional extension of skin. *Science* 215:174–176

Zeigler FC, Landeen L, Naughton GK, et al (1993): Tissue-engineered, three-dimensional human dermis to study extracellular matrix formation in wound healing. *J Toxicol-Cut Ocular Toxicol* 12:303–312

8

TISSUE ENGINEERING IN THE GENITOURINARY SYSTEM

NTHONY ATALA, M.D.

INTRODUCTION

In general, genitourinary tissue has traditionally been replaced with autologous tissues harvested from other sites of the body (Atala and Retik, 1994). However, these approaches rarely replace the entire function of the original tissue. Congenital defects occur more often in the genitourinary system than in any other system within the human body. The genitourinary system, composed of the kidneys, ureters, bladder, urethra, and genital organs, is exposed to a variety of possible injury sites as the fetus develops. Some of the congenital abnormalities require early reconstruction with nonurologic tissues. For example, a baby born with bladder exstrophy, a condition wherein the bladder develops outside the abdominal wall, requires prompt surgical intervention and possibly reconstruction of the bladder with nonurologic tissues, such as gastrointestinal segments. Similarly, baby girls born with vaginal maldevelopment, may require reconstruction with intestinal segments or skin (Atala and Hendren, 1994; Hendren and Atala, 1994). Boys born with hypospadias, a condition wherein the urethra does not develop normally, require early creation of a neo-urethra with skin or mucosa from multiple body sites (Atala and Retik, 1996; Retik et al, 1994).

The genitourinary tract may also be affected by other nonurologic conditions. For example, individuals born with myelodysplasia, or occult dysraphisms, which represent defects within the spinal cord, may have severe bladder dysfunction (Atala et al, 1992a). Some of the patients require bladder reconstruction with gastrointestinal segments (Atala et al, 1993a).

Synthetic Biodegradable Polymer Scaffolds
Anthony Atala and David Mooney, Editors
Robert Langer & Joseph P. Vacanti, Associate Editors
© 1997 Birkhäuser Boston

Aside from congenital abnormalities, individuals may also suffer from acquired disorders such as cancer. Bladder cancer may require removal of the diseased organ and creation of a neo-bladder. Gastrointestinal tissue is usually used for this application. Other conditions, such as trauma, may lead to genitourinary organ damage or loss, requiring eventual reconstruction.

The replacement of kidney function is yet an even more difficult challenge. Millions of patients worldwide suffer from kidney failure, requiring long-term dialysis. The majority of these patients would benefit from kidney transplantation but are unable to proceed with that mode of treatment due to a severe organ shortage. Even if kidney transplantation is achieved, patients require permanent immunosuppressive therapy and may have multiple life-threatening complications.

A large number of materials, including naturally-derived materials and synthetic polymers, have been utilized to fabricate synthetic prosthesis for the genitourinary system. The most common type of synthetic prostheses for urologic use are made of silicone. Silicone prostheses have been used for the treatment of urinary incontinence (artificial urinary sphincter) (Bauer et al, 1986), vesicoureteral reflux (detachable balloon system, silicone microparticles) (Atala et al, 1992b; Buckley et al, 1991), and impotence (penile prostheses) (Riehmann et al, 1993). The literature is replete with publications reporting the wide array of complications (i.e., device malfunction, infection, etc) associated with these devices.

In some disease states, such as urinary incontinence or vesicoureteral reflux, artificial agents (Teflon paste, silicone microparticles) have been used as injectable bulking substances; however, these substances are not entirely biocompatible (Atala, 1994).

The use of nonurologic tissues in the genitourinary tract is common due to a lack of a better alternative, despite the known possible adverse effects. For example, the use of bowel in genitourinary reconstruction is associated with a variety of complications (Atala and Retik, 1994). These include metabolic abnormalities, infection, perforation, urolithiasis, increased mucous production, and malignancy. Alternative approaches need to be developed to overcome the problems associated with the incorporation of intestinal segments into the urinary tract. Natural tissues and synthetic materials that have been tried previously in experimental and clinical settings include omentum, peritoneum, seromuscular grafts, de-epithelialized segments of bowel, polyvinyl sponge, and polytetrafluoroethylene (Teflon) (Atala and Retik, 1994). These attempts have usually failed. It is self-evident that urothelial to urothelial anastomoses are preferable functionally. However, the limited amount of autologous urothelial tissue for reconstruction generally precludes this option. There is a critical need for tissues to replace lost and functionally-deficient genitourinary tissues (Atala, 1995). Engineering tissues using selective cell transplantation may provide a means to create functional new genitourinary tissues (Langer and Vacanti, 1993).

In tissue engineering, donor tissue is dissociated into individual cells or small tissue fragments and either implanted directly into the autologous host or attached to a support matrix, expanded in culture, and reimplanted after expansion. Ideally, this approach might allow lost tissue function to be restored or replaced in toto and with limited complications.

The success of using cell transplantation strategies for genitourinary reconstruction depends on the ability to use donor tissue efficiently and to provide the right conditions for long term survival, differentiation, and growth. We have achieved an approach to urologic tissue regeneration by patching isolated cells to a support structure that would have suitable surface chemistry for guiding the reorganization and growth of the cells. (Atala et al, 1992c; Atala et al, 1993c). The supporting matrix is composed of crossing filaments which can allow cell survival by diffusion of nutrients across short distances once the cell-support matrix is implanted. Ideally, the cell-support matrix would become vascularized in concert with expansion of the cell mass following implantation.

CELL GROWTH AND REGULATION

One of the initial limitations of applying tissue engineering techniques to the urinary tract has been the previously encountered inherent difficulty of growing genitourinary associated cells in large quantities. Even as recently as 6 years ago, it was believed that urothelial cells, which line most of the urinary tract, had a natural senescence that was hard to overcome. Normal urothelial cells could be grown in the laboratory setting, but only for a limited period of time of a few weeks, and without any demonstrable expansion.

We have recently shown that normal human bladder epithelial cells can be efficiently harvested from surgical material, extensively expanded in culture in serum-free conditions, and their differentiation characteristics, growth requirements, and other biological properties studied (Cilento et al, 1994). Using our methods of cell culture we estimate that it would be theoretically possible to expand a urothelial strain from a single specimen that initially covers a surface area of 1 cm^2 to one covering a surface area of 4202 m^2 (the equivalent area of one football field) within 8 weeks even if it were assumed that 50% of the cells would be lost with each passage. This indicates that it should be possible to collect autologous urothelial cells from human patients, expand them in culture, and return them to the human donor in sufficient quantities for reconstructive purposes. Human bladder smooth muscle can be isolated in a similar fashion and studied transiently under serum-free conditions. The ability to greatly expand primary populations of human urothelial-associated cells in serum-free conditions is important because it indicates that frequent or continuous replenishment of in vitro stocks of normal cells from fresh surgical material, a requirement that would significantly inhibit the use of these cells for research purposes, is not needed.

Whether the cells retain a normal program of differentiation is an issue of great importance. In our studies, immunofluorescent cell staining with broadly-reacting anticytokeratin antibodies (AE1/AE3) and an antibody that reacts specifically to cytokeratin 7, present in bladder mucosa in vivo, confirmed the epithelial phenotype of the cells and the cells' bladder origin. The epithelial cells also stained positively with an antibody to vimentin, suggesting that the cells did not exhibit a barrier-forming epithelial phenotype. To examine this possibility further, the urothelial cells were stained with an antibody to E-cadherin, a ubiquitous epithelial cell surface adhesion protein. The barrier-forming kidney cell line MDCK, which forms intercellular epithelial adhesion complexes, stained positively with this antibody; however, the urothelial cells stained very weakly. This suggested that the cells lacked a ubiquitous protein found in adherens junctions in barrier-forming epithelia. Northern blot hybridization with an anti-E-cadherin cDNA probe confirmed that the urothelial cells expressed low levels of E-cadherin mRNA when compared to a barrier-forming prostate epithelial cell line. These results indicate that the urothelial cells grown on plastic do not form functional epithelial monolayers under these culture conditions. Urothelial cells grown on collagen type I matrix were positive for the expression of E-cadherin as detected by indirect immunofluorescence. Cells grown on Matrigel surfaces formed complex cell aggregates in sheet- and tubelike structures. Confocal microscopy demonstrated that cells grown on Matrigel localized E-cadherin to interepithelial junctions. Therefore epithelial differentiation of these cells can be induced by some extracellular matrices (Tobin et al, 1994). Also of concern is whether the cells retain a normal cytogenetic component. Chromosomal analysis was performed on urothelial cells at the third and sixth passages. A normal chromosomal complement with no detectable abnormalities was observed by Giemsa staining of metaphase chromosomes.

Most of the growth factor biology in human urothelial-associated cells is entirely unknown. This area of study is important in order to further understand the regulatory mechanisms involved with the growth and development of these cells and how they ultimately affect the entire genitourinary system. Studies performed at our laboratory have demonstrated that unlike keratinocytes and a variety of other epithelia, human bladder urothelial cells do not require EGF for growth. Growth kinetics using multiple independently-derived urothelial strains were observed to be essentially identical under (+)EGF and (−)EGF conditions. These data highlight an important potential difference between the keratinocyte and bladder in vitro epithelial cell models. One possible explanation for these observations is that EGF may not play a role in normal bladder epithelial cell growth. However, epithelial cells in the germinal layer of the bladder epithelium, which are thought to repopulate the suprabasal epithelial layers after injury, express EGF receptors (Messing et al, 1987). Despite the apparent absence of a growth-promoting effect of EGF, these observations suggest a role for a regulatory pathway involving the EGF receptor in urothelial cell growth and/or differentiation. Recently it has become evident that autocrine or paracrine regulation through the EGF receptor does not require binding of the receptor to EGF. Multiple high-affinity ligands for the EGF receptor, which are capable of activating EGF receptor-mediated signal transduction pathways in a fashion similar to EGF, have recently been identified (Prigent, 1992). At present these include TGFa, amphiregulin, heparin binding EGF-like growth factor, and betacellulin. The discovery of these distinct EGF-like molecules indicates that the EGF receptor is a degenerate receptor capable of binding different activating ligands with high specificity in different contexts. We have shown that human bladder urothelial cells constitutively express high levels of mRNAs encoding three EGF receptor ligands: HB-EGF, TGFa, and amphiregulin (Atala et al, 1996; Freeman et al, 1995a,b). EGF transcripts were not detected. These results suggest that normal bladder epithelia might in fact utilize EGF receptor pathways during regulated growth. Expression of alternative EGF receptor ligands by transitional epithelia might explain the apparent EGF-independent growth of these cells. HB-EGF mRNA levels were highly inducible by phorbol ester treatment, while AR mRNA was not responsive to phorbol ester, suggesting that HB-EGF and AR are regulated dissimilarly at the genetic level in urothelial cells.

CELL DELIVERY VEHICLES

The support structure chosen for cell delivery is of utmost importance. It is known from previous studies that artificial permanent support structures are lithogenic (Teflon, silicone). Other investigators have tried permanent homograft or heterograft support structures such as dura; however, these contract with time and are problematic in a clinical setting. Natural permanent support structures, such as denuded bowel, retain their inherent properties and mucosal regrowth invariably occurs with time. A variety of synthetic polymers, both degradable and nondegradable, have been utilized to fabricate tissue engineering matrices (Langer and Vacanti, 1993; Peppas and Langer, 1994; Tachibana, et al, 1985; Thüroff et al, 1983). One of the earliest reported studies in the entire tissue engineering field was an attempt to line biodegradable tubular structures with cultured smooth muscle cells obtained from bladder wall biopsies (Langer, 1990). These constructs were proposed to have wide applicability in genitourinary operations. Another early attempt to engineer urologic structures involved implanting collagen sponge tubes as ureteral replacements (Mooney et al, 1994). It was hoped that the implanted matrix would induce migration of epithelial cells from the adjacent tissue and the formation of an epithelial cell-lined tubular tissue. However, salt deposits onto the collagen matrix were noted following exposure to urine.

Synthetic polymers can be manufactured reproducibly and can be designed to exhibit the necessary mechanical properties (Langer, 1990). Among synthetic materials, resorbable polymers are preferable because permanent polymers carry the risk of infection, calcification, and unfavorable connective tissue response. Polymers of lactic and glycolic acid have been extensively utilized to fabricate tissue engineering matrices (Atala et al, 1992c; 1993c; Langer and Vacanti, 1993; Mooney et al, 1994; 1996a,c,d). These polymers have many desirable features; they are biocompatible, processable, and biodegradable. Degradation occurs by hydrolysis, and the time sequence can be varied from weeks to over a year by manipulating the ratio of monomers and by varying the processing conditions. These polymers can be readily formed into a variety of structures, including small diameter fibers and porous films.

The porosity, pore size distribution, and continuity dictate the interaction of porous materials and transplanted cells with the host tissue. Fibrovascular tissue will invade a device if the pores are larger than approximately 10 mm, and the rate of invasion will increase with the pore size and total porosity of a device (Mikos et al, 1993; Weslowski et al, 1961; White et al, 1981). This process results in the formation of a capillary network in the developing tissue (Mikos et al, 1993). Vascularization of the engineered tissue may be required to meet the metabolic requirements of the tissue and integrate it with the surrounding host. In urologic applications it may also be desirable to have a nonporous luminal surface (e.g., to prevent leakage of urine from the tissue) (Olsen et al, 1992).

ENGINEERING GENITOURINARY TISSUES: SURGICAL IMPLANTS

The direction that we have followed to engineer urologic tissue involves the use of biodegradable materials that act as cell delivery vehicles (Langer, 1990). Therefore, we have focused our efforts on urothelial and bladder muscle cell transplantation using this cell-biodegradable matrix model for genito-urinary reconstructive applications.

We have carried out a series of in vivo cell-polymer experiments. Histologic analysis of human urothelial, bladder muscle, and composite urothelial and bladder muscle-polymer scaffolds, implanted in athymic mice and retrieved at different time points, indicated that viable cells were evident in all three experimental groups (Atala et al, 1993c). Implanted cells oriented themselves spatially along the polymer surfaces. The cell populations appeared to expand from one layer to several layers of thickness with progressive cell organization with extended implantation times. Polymers alone evoked an angiogenic response by 5 days, which increased with time. Polymer fiber degradation was evident after 20 days. An inflammatory response was also evident at 5 days, and its resolution correlated with the biodegradation sequence. Cell-polymer composite implants of urothelial and muscle cells, retrieved at extended times (50 days), showed extensive formation of multilayered sheetlike structures and well-defined muscle layers. Polymers seeded with cells and manipulated into a tubular configuration showed layers of muscle cells lining the multilayered epithelial sheets. Cellular debris appeared reproducibly in the luminal spaces, suggesting that epithelial cells lining the lumina are sloughed into the luminal space. Cell polymers implanted with human bladder muscle cells alone showed almost complete replacement of the polymer with sheets of smooth muscle at 50 days. This experiment demonstrated, for the first time, that composite tissue engineered structures could be created de novo. Prior to this study, only single cell type tissue engineered structures had been created.

The cells delivered in vivo retained a normal program of differentiation. Western blot analysis using broadly-reactive anticytokeratin antibodies demonstrated the presence of a 40kDa cytokeratin in protein fractions isolated from polymers which comigrated with a strongly immunoreactive band seen in freshly isolated urothelial cells. Cytokeratin 7, a urothelial-associated cytokeratin, was also identifiable in the retrieved specimens. In addition, urothelial cells synthesizing DNA were identified using the bromodeoxyuridine (BrdU) cell labeling method. Anti-BrdU detection of metabolically labeled DNA suggested that the human cells proliferated on the polymers.

This approach has recently been expanded to engineer new functional urologic structures (Cilento et al, 1995; Yoo et al, 1995; 1996). In one study conducted in dogs, urothelial and smooth muscle cells were harvested, expanded in vitro, and seeded onto biodegradable polymer scaffolds. These structures were tubularized and used to replace ureteral segments in each animal (Yoo et al, 1995). Beagle bladder tissue specimens were microdissected, and the mucosal and muscular layers separated. Both urothelial and smooth muscle cells were harvested and expanded separately. Cells were seeded onto preformed tubular polyglycolic acid polymers in vitro. Cell-polymer scaffolds were created consisting of urothelial and smooth muscle, urothelial, and smooth muscle cell population(s). Polymers without cells served as controls. Beagles underwent partial ureterectomies, and the polymer scaffolds were used for ureteral replacement in each animal. Double-J catheters were used as ureteral stents. Four to six weeks after ureteroplasty, the Double-J stents were removed, and intravenous pyelography was performed in all dogs to confirm the patency of the neo-ureteral segments. All implanted polymer scaffolds were retrieved at six weeks. Polymer, polymer-urothelial cell, bladder muscle, and urothelial/bladder muscle implants were found to contain a normal cellular organization consisting of a urothelial lined lumen surrounded by overlying submucosal tissue and a layer of smooth muscle. An angiogenic response and polymer fiber degradation was evident in all animals. These results suggest that the creation of artificial ureters may be achieved in vivo using biodegradable polymers as transplanted cell delivery and native cell expansion vehicles. The malleability of the synthetic polymer allowed for the creation of cell-polymer implants manipulated into preformed tubular configurations. The combination of both smooth muscle and urothelial cell-polymer scaffolds is able to provide a template wherein a functional ureter may be created de novo.

In other sets of experiments, the same approach was used to augment bladders (Yoo et al, 1996). Beagle bladder tissue specimens were microdissected, and the mucosal and muscular layers separated. Both urothelial and smooth muscle cells were harvested and expanded separately. The cells were seeded onto 5×5 and 10×10 cm sized sheets of polyglycolic acid polymers in vitro. Cell-polymer scaffolds were created consisting of urothelial and smooth muscle, urothelial, and smooth muscle cell populations. Polymers without cells served as controls. Beagles underwent cruciate cystotomies on the bladder dome, and the polymer scaffolds were used to augment the bladder in each animal. Omentum was wrapped over the augmented bladder to enhance angiogenesis to the polymer-cell complex. Cystostomy catheters were inserted to decompress the bladder and were removed after 10–14 days. Urodynamic studies and fluoroscopic cystography were performed in all dogs pre- and postoperatively. The augmented bladders were retrieved at 6, 8, and 12 weeks. The average increase in bladder capacity was approximately 40%. Furthermore, the average compliance of the augmented bladder increased an average of 30%. Polymer, polymer-urothelial cell, bladder muscle, and urothelial/bladder muscle implants were found to contain a normal cellular organization consisting of a urothelial lined lumen surrounded by overlying submucosal tissue and ingrowth of smooth muscle. An angiogenic response and polymer fiber degradation were evident in all animals. These results show that the creation of artificial functional and anatomical bladders may be achieved in vivo; however, much work remains to be done in terms of the functional parameters of these implants. The same strategy has been used in trying to achieve urethral reconstruction. (Cilento et al, 1995).

The kidney is responsible not only for urine excretion but for several other important metabolic functions in which critical kidney by-products, such as renin, erythropoietin, and Vitamin D play a large role. End stage renal failure is a devastating disease involving multiple organs in affected individuals. Although dialysis can prolong survival for many patients with end stage renal disease, only renal transplantation can currently restore normal function. Renal transplantation is severely limited by a critical donor shortage. Augmentation of either isolated or total renal function with kidney cell expansion in vitro and subsequent autologous transplantation may be a feasible solution. Toward this goal we explored the feasibility of achieving renal cell growth and expansion in vitro with subsequent implantation (Atala et al, 1995).

New Zealand white rabbits underwent nephrectomy and renal artery perfusion with a nonoxide solution that promoted iron particle entrapment in the glomeruli. Homogenization of the renal cortex and fractionation in 83 and 210 micron sieves with subsequent magnetic extraction yielded three separate purified suspensions of distal tubules, glomeruli, and proximal tubules. These cells were plated separately in vitro and seeded onto biodegradable polyglycolic acid polymer scaffolds. Polymer scaffolds were implanted subcutaneously into host athymic mice. This included implants of proximal tubular cells, glomeruli, distal tubular cells, and a mixture of all three cell types. Cell-polymer scaffolds were implanted as sheets in a flat configuration. Polymers alone served as controls. Animals were sacrificed at one week, two weeks, and one month after implantation, and the retrieved scaffolds were examined histologically. An acute inflammatory phase and a chronic foreign body reaction were seen, accompanied by vascular ingrowth, by 7 days after implantation. Histologic examination demonstrated progressive formation and organization of the nephron segments within the polymer fibers with time. Renal cell proliferation in the implanted cell-polymer scaffolds was detected by in vivo labeling of replicating cells with the thymidine analog bromodeoxyuridine (BrdU). BrdU incorporation into renal cell DNA was identified immunocytochemically with a monoclonal anti-BrdU antibody.

These results demonstrate that renal specific cells can be successfully harvested, will survive in culture, and will attach to artificial biodegradable polymers. The renal cell-polymer scaffolds can be implanted into host animals where the cells replicate and organize into nephron segments as the polymer, which acts as a cell delivery vehicle, undergoes biodegradation. These findings suggest that it may be possible to use in vitro expanded autologous kidney cells in combination with biodegradable polymers for either an isolated or general gain in renal function (Atala et al, 1995).

ENGINEERING GENITOURINARY TISSUES: INJECTABLE THERAPIES

Both urinary incontinence and vesicoureteral reflux are common conditions affecting the genitourinary system wherein injectable bulking agents can be used for treatment. Urinary incontinence may result from a deficient or weak musculature in the bladder neck and urethral area. Primary vesicoureteral reflux results from a congenitally deficient longitudinal submucosal muscle of the distal ureter which leads to an abnormal flow of urine from the bladder to the upper tract. The anatomical features that characterize the normal mechanism of the ureterovesical junction are: (1) an oblique entry of the ureter into the bladder, (2) an adequate intramural ureter, and (3) support of the detrusor muscle. In vesicoureteral reflux, these factors are inadequate or absent, and reflux results (Atala and Casale, 1990). The diagnosis of reflux requires invasive radiographic studies, although less invasive diagnostic modalities are being developed (Atala et al, 1993e; 1994c). All of the surgical techniques for the correction of incontinence and vesicoureteral reflux attempt to achieve a normal anatomy.

Although open surgical procedures for the correction of urinary incontinence and reflux have excellent results in the hands of experienced surgeons, it is associated with a well-recognized morbidity, including the pain and immobilization of a lower abdominal incision, bladder spasms, hematuria, and postoperative voiding frequency. In an effort to avoid open surgical intervention, widespread interest was initiated by Berg's clinical experience with the endoscopic injection of Teflon paste in 1973. (Berg, 1973). With this technique, a cystoscope is inserted into the bladder, a needle is inserted through the cystoscope and placed under direct vision in the submucosal space, and Teflon paste is injected (Politano, 1982). The teflon paste, injected endoscopically, treats the incontinence by acting as a bulking material which increases urethral resistance. Vesicoureteral reflux was treated in a similar manner, by injecting Teflon in the subureteral space (O'Donnell and Puri, 1984). However, soon after the introduction of this treatment, a controversy regarding the use of Teflon paste ensued. Teflon particles injected in animals were noted to migrate to distant organs, such as the lungs, liver, spleen, and brain, inducing granuloma formation (Malizia et al, 1984). Polytetrafluoroethylene migration and granuloma formation have also been reported in humans (Claes et al, 1989). Teflon's safety for human use was questioned, and the paste was thereafter not approved by the FDA. Silicone microparticles have been used as bulking agent for the treatment of urinary incontinence and vesicoureteral reflux (Buckley et al, 1991). However, silicone particles have been shown to migrate to distant organs (Henly et al, 1995). Collagen injections have been used in a similar manner. (Leonard et al, 1991). However, collagen loses its volume over time, leading to treatment failure in the majority of patients (Atala, 1994).

There are definite advantages in treating urinary incontinence and vesicoureteral reflux endoscopically. The method is simple and can be completed in less than 15 minutes, it has a low morbidity, and it can be performed on an outpatient basis. The goal of several investigators has been to find alternate implant materials that would be safe for human use.

Laparoscopic approaches for incontinence and reflux have been attempted and are technically feasible (Atala, 1993; Atala et al, 1993d). However, at least two surgeons with laparoscopic expertise are needed, the length of the procedure is much longer than with open surgery, the surgery is converted from an extraperitoneal to an intraperitoneal approach, and the cost is higher due to both increased operation time and the expense of the disposable laparoscopic equipment.

Despite the fact that over a decade has transpired since the Teflon controversy, little progress has been made in this area of research. The ideal substance for the endoscopic treatment of reflux should be injectable, nonantigenic, nonmigratory, volume stable, and safe for human use. Toward this goal we had previously conducted long-term studies to determine the effect of injectable chondrocytes in vivo. (Atala et al, 1993b). We initially determined that alginate, a simple sugar solution embedded with chondrocytes, would serve as a synthetic substrate for the injectable delivery and maintenance of cartilage architecture in vivo. The use of autologous cartilage for the treatment of vesicoureteral reflux in humans would satisfy all the requirements for an ideal injectable substance. A biopsy of the ear could be easily and quickly performed (analogous to ear piercing) followed by chondrocyte processing and endoscopic injection of the autologous chondrocyte suspension for the treatment reflux.

Chondrocytes can be readily grown and expanded in culture. Neocartilage formation can be achieved in vitro and in vivo using chondrocytes cultured on synthetic biodegradable polymers (Atala et al, 1993b). We used alginate, a liquid solution of gluronic and mannuronic acid, as the chondrocyte delivery vehicle. Alginate undergoes hydrolytic biodegradation; its degradation time can be varied depending on the concentration of each of the polysaccharides. In our experiments, the cartilage matrix replaced the alginate as the polysaccharide polymer underwent biodegradation. We then adapted the system for the treatment of vesicoureteral reflux in a porcine model (Atala et al, 1994b).

Six mini-swine underwent bilateral creation of reflux. All six were found to have bilateral reflux without evidence of obstruction at three months following the procedure. Chondrocytes were harvested from the left auricular surface of each min-swine and expanded with a final concentration of $50–150 \times 10^6$ viable cells per animal. The animals then underwent endoscopic repair of reflux with the injectable autologous chondrocyte solution on the right side only.

Cystoscopic and radiographic examinations were performed at two, four, and six months after treatment. Cystoscopic examinations showed a smooth bladder wall. Cystograms showed no evidence of reflux on the treated side and persistent reflux in the uncorrected control ureter in all animals. All animals had a successful cure of reflux in the repaired ureter without evidence of hydronephrosis on excretory urography. The harvested ears had evidence of cartilage regrowth within one month of chondrocyte retrieval.

At the time of sacrifice, gross examination of the bladder injection site showed a well-defined rubbery to hard cartilage structure in the subureteral region. Histologic examination of these specimens using hematoxylin and eosin stains showed evidence of normal cartilage formation. The polymer gels were progressively replaced by cartilage with increasing time. Aldehyde fuschin-alcian blue staining suggested the presence of chondroitin sulfate. Microscopic analyses of the tissues surrounding the injection site showed no inflammation. Tissue sections from the bladder, ureters, lymph nodes, kidneys, lungs, liver, and spleen showed no evidence of chondrocyte or alginate migration, or granuloma formation.

Our studies showed that chondrocytes can be easily harvested and combined with alginate in vitro, the suspension can be easily injected cystoscopically and the elastic cartilage tissue formed is able to correct vesicoureteral reflux without any evidence of obstruction. (Atala et al, 1994b).

Using the same line of reasoning as with the chondrocyte technology, our group investigated the possibility of using autologous muscle cells. (Atala et al, 1994a). Using athymic nu/nu mice, an experiment was conducted to investigate whether muscle cells could be used for the endoscopic treatment of reflux. This study demonstrated similar results when compared to the chondrocyte-alginate gel suspension experiments. After injection of the muscle cell-alginate complex, the alginate was progressively replaced with muscle cells.

Given these encouraging results, in vivo experiments were conducted in five Hanford mini-pigs to see if reflux could be corrected. (Atala et al, 1994d). Muscle cells were harvested and plated in vitro. After expansion, the cells were individually quantitated and concentrated to 20 million cells per cc. The cell suspensions were complexed with sodium alginate and calcium sulfate. A cystoscopic 21-gauge needle was used to inject the muscle suspension into a subureteric region of the refluxing ureter. Cystograms demonstrated 100% correction of reflux on the treated side and persistence of reflux on the contralateral untreated side. There was no evidence of obstruction on the treated side by excretory urography. Histological examination showed that the injected muscle cells survive in the subureteric region. The cell nidus corrected reflux and was stable, biocompatible, nonantigenic, and nonmigratory.

In addition to its use for the endoscopic treatment of reflux and urinary incontinence, the system of injectable autologous cells may also be applicable for the treatment of other medical conditions, such as rectal incontinence, dysphonia, plastic reconstruction, and wherever an injectable permanent biocompatible material is needed.

FUTURE DIRECTIONS

Most of the effort expended to engineer genitourinary tissues has occurred within the last 5 years. Recent successes suggest that engineered urologic tissues may have clinical utility in the near future.

A tremendous effort is being expanded to develop biologically active synthetic materials that can be utilized to fabricate tissue engineering matrices (Hubbell, 1995; Barrera et al, 1993; Peppas and Langer, 1994). These materials will likely find applications in engineering genitourinary tissues. A critical feature of many genitourinary tissues is the specific organization of different cell types within the tissue. The studies described above suggest that certain transplanted cells have an inherent ability to reorganize into appropriate structures following implantation. However, this may not be true of other cell types. In either case, it may be possible to control the organization of different cell types within an engineered tissue through the development of matrices that specifically promote the adhesion of certain cell types. For example, a short peptide has been identified that promotes endothelial cell adhesion but not fibroblast or smooth muscle adhesion. (Hubbell et al, 1991). A significant effort is also being expended to determine the role of specific growth factors in the process of tissue regeneration. The identification of factors capable of directing tissue development, and the development of techniques to deliver these factors to transplanted cells may hasten the engineering of clinically relevant genitourinary tissues (Atala et al, 1996; Freeman et al, 1995a,b; Mooney et al, 1996b).

References

Atala A (1993): Laparoscopic technique for the extravesical correction of vesicoureteral reflux. *Dial Ped Urol* 16:12

Atala A (1994): Use of non-autologous substances in VUR and incontinence treatment. *Dial Ped Urol* 17:11–12

Atala A (1995): Tissue engineering in the urinary tract. *Dial Ped Urol* 18:1

Atala A, Casale, AJ (1990): Management of primary vesicoureteral reflux. *Inf Urol* 2:39–42

Atala A, Hendren WH (1994): Reconstruction with bowel segments. *Dial Ped Urol* 17:7

Atala A, Retik A (1994): Pediatric urology—future perspectives. In: *Clinical Urology*, Krane RJ, Siroky MB, Fitzpatrick JM, ed. Philadelphia: J.B. Lippincott

Atala A, Retik AB (1996): Hypospadias. In: *Reconstructive Urologic Surgery*, Libertino JA, Zinman L, eds. Baltimore: The Williams and Wilkins Co

Atala A, Bauer SB, Dyro FM, Shefner J, Shillito J, Sumeer S, Scott MR (1992a): Bladder functional changes resulting from lipomyelomeningocele repair. *J Urol* 148:592

Atala A, Peters CA, Retik AB, Mandell J (1992b): Endoscopic treatment of vesi-coureteral reflux with a self-detachable balloon system. *J Urol* 148:724–728

Atala A, Vacanti JP, Peters CA, Mandell J, Retik AB, Freeman MR (1992c): Forma-tion of urothelial structures in vivo from dissociated cells attached to biodegrad-able polymer scaffolds in vitro. *J Urol* 148:658–662

Atala A, Bauer SB, Hendren WH, Retik AB (1993a): The effect of gastric augmen-tation on bladder function. *J Urol* 149:1099

Atala A, Cima LG, Kim W, Paige KT, Vacanti JP, Retik AB, Vacanti CA (1993b): Injectable alginate seeded with chondrocytes as a potential treatment for vesi-coureteral reflux. *J Urol* 150:745–747

Atala A, Freeman MR, Vacanti JP, Shepard J, Retik AB (1993c): Implantation in vivo and retrieval of artificial structures consisting of rabbit and human urothelium and human bladder muscle. *J Urol* 150:608–612

Atala A, Kavoussi LR, Goldstein DS, Retik AR, Peters CA (1993d): Laparoscopic correction of vesicouretral reflux. *J Urol* 150:748

Atala A, Wible JH, Share JC, Carr MC, Retik AB, Mandell J (1993e): Sonography with sonicated albumin in the detection of vesicoureteral reflux. *J Urol* 150:756–758

Atala A, Cilento BG, Paige KT, Retik AB (1994a): Injectable alginate seeded with human bladder muscle cells as a potential treatment for vesicoureteral reflux. *J Urol* 151(suppl):5

Atala A, Kim W, Paige KT, Vacanti CA, Retik AB (1994b): Endoscopic treatment of vesicoureteral reflux with chondrocyte-alginate suspension. *J Urol* 152:641–643

Atala A, Share JC, Paltiel HJ, Grant R, Retik A (1994c): Sonicated albumin in the detection of vesicoureteral reflux in humans. *Soc Ped Urol Newsl* 8:6–7

Atala A, Shepard JA, Retik AB (1994d): Endoscopic treatment of reflux with autolo-gous bladder muscle cells. In: *Proceedings, Pediatric Urology Section, American Academy of Pediatrics.* Dallas, TX

Atala A, Schlussel RN, Retik AB (1995): Renal cell growth in vivo after attachment to biodegradable polymer scaffolds. *J Urol* 153 (suppl):4

Atala A, Yoo J, Raab G, Klagsburn M, Freeman MR (1996): Regulated secretion of an engineered growth factor by human urothelial cells in primary culture. *J Urol* 155(suppl):5

Barrera DA, Zylstra E, Lansbury PT, et al. Synthesis and RGD peptide modification of a new biodegradable copolymer: Poly(lactic acid-co-lysine). *J Am Chem Soc* 115:11010–11011

Bauer SB, Reda EF, Colodny AH, Retik AR (1986): Detrusor instability: A delayed complication in association with the artificial sphincter. *J Urol* 135: 1212

Berg S (1973): Urethroplastie par injection de polytef. *Arch Surg* 107:379.21

Buckley JF, Scott R, Aitchison M, et al. (1991): Periurethral microparticulate silicone injection for stress incontinence and vesicoureteric reflux. *Min Inv Ther* 1 (suppl):72

Cilento BJ, Freeman MR, Schneck FX, Retik AB, Atala AN (1994): Phenotypic and cytogenetic characterization of human bladder urothelia expanded in vitro. *J Urol* 152:665–670

Cilento BG, Retik AB, Atala A (1995): Urethral reconstruction using a polymer mesh. *J Urol* 153(suppl):4

Claes H, Stroobants D, Van Meerbeek J, Verbeken E, Knockaert D, Beart L (1989):

Pulmonary migration following periurethral polytetrafluoroethylene injection for urinary incontinence. *J Urol* 142:821–822

Freeman MR, Schneck FX, Soker S, Raab G, Tobin M, Yoo J, Klagsbrun M, Atala A (1995a): Human urothelial cells secrete and are regulated by heparin-binding epidermal growth factor-like growth factor (HB-EGF). *J Urol* 153(suppl):4

Freeman MR, Schneck FX, Klagsbrun M, Atala A (1995b): Growth factor biology of human urothelial cells grown under serum-free conditions. *J Urol* 153(suppl):4

Hendren WH, Atala A (1994): Use of bowel for vaginal reconstruction. *J Urol* 152:752–755

Henly DR, Barrett DM, Welland TL, O'Connor MK, Malizia AA, Wein AJ (1995): Particulate silicone for use in periurethral injections: Local tissue effects and search for migration. *J Urol* 153:2039–2043

Hubbell JA, (1995): Biomaterials in tissue engineering. *Bio/Technology* 13:565–576

Hubbell JA, Massia SP, Desai NP (1991): et al. Endothelial cell-selective materials for tissue engineering in the vascular graft via a new receptor. *Bio/Technology* 9:568–572

Langer R (1990): New methods of drug delivery. *Science* 249:1527

Langer R, Vacanti JP (1993): Tissue engineering. *Science* 260:920–926

Leonard MP, Canning DA, Peters CA, Gearhart JP, Jeffs RD (1991): Endoscopic injection of glutaraldehyde cross-linked bovine dermal collagen for correction of vesicoureteral reflux. *J Urol* 145:115–119

Malizia AA, Reiman HM, Myers RP, Sande JR, Barham SS, Benson RC, Dewanjee MK, Utz WJ (1984): Migration and granulomatous reaction after periurethral injection of polytef (Teflon). *JAMA* 251:3277–3281

Messing EM, Hanson P, Ulrich P, Erturk E (1987): Epidermal growth factor—Interactions with normal and malignant urothelium: In vivo and in situ studies. *J Urol* 138:1329

Mikos AG, Sarakinos G, Lyman MD (1993): et al. Prevascularization of porous biodegradable polymers. *Biotechnol Bioeng* 42:716–723

Mooney DJ, Organ G, Vacanti JP, Langer R (1994): Design and fabrication of biodegradable polymer devices to engineer tubular tissues. *Cell Transplant* 3:438–446

Mooney DJ, Breuer C, McNamara K (1996a): et al. Fabricating tubular tissues with devices of poly(D,L-lactic-co-glycolic acid). *Tissue Eng:* in press

Mooney D, Kaufmann PM, Sano S, McNamara K, Schwendeman S, Vacanti JP, Langer R (1996b): Localized delivery of epidermal growth factor improves the survival of transplanted hepatocytes. Submitted

Mooney DJ, Mazzoni CL, Breuer C, McNamara K, Hern D, Vacanti JP, Langer R (1996c): Stabilized polyglycolic acid fiber-based devices for tissue engineering. *Biomaterials:* in press

Mooney D, Park S, Kaufmann PM, Sano S, McNamara K, Vacanti JP, Langer R (1996d): Biodegradable sponges for hepatocyte transplantation. *J Biomed Mat Res:* in press

O'Donnell B, Puri P (1984): Treatment of vesicoureteric reflux by endoscopic injection of Teflon. *Br Med J* 289:7–9

Olsen L, Bowald S, Busch C, Carlsten J, Eriksson I (1992): Urethral reconstruction with a new synthetic absorbable device. *Scan J Urol Nephrol* 26:323–326

Politano VA (1982): Periurethral polytetrafluoroethylene injection for urinary incontinence. *J Urol* 127:439–442

Prigent SA (1992): The type 1 (EGFR-related) family of growth factor receptors and their ligands. *Prog Growth Factor Res* 4:1–24

Retik AB, Bauer SB, Mandell J, Peters CA, Colodny A, Atala A (1994): Management of severe hypospadias with a 2-stage repair. *J Urol* 152:749–751

Riehmann M, Gasser TC, Bruskewitz RC (1993): The hydroflex penile prosthesis: A test case for the introduction of new urological technology. *J Urol* 149:1304–1307

Tachibana M, Nagamatsu GR, Addonizio JC (1985): Ureteral replacement using collagen sponge tube grafts. *J Urol* 133:866–869

Thüroff JW, Bazeed MA, Schmidt RA, Luu DJ, Tanagho EA (1983): Cultured rabbit vesical smooth muscle cells for lining of dissolvable synthetic prosthesis. *Urol* 21:155–158

Tobin MS, Freeman MR, Atala A (1994): Maturational response of normal human urothelial cells in culture is dependent on extracellular matrix and serum additives. *Surg For* 45:786

Wesolowski SA, Fries CC, Karlson KE (1961): et al. Porosity: Primary determinant of ultimate fate of synthetic vascular grafts. *Surg* 50:91–96

White RA, Hirose FM, Sproat RW et al. Histopathologic observations after short-term implantation of two porous elastomers in dogs. *Biomaterials* 2:171–176

Yoo J, Satar N, Retik AB, Atala A (1995): Ureteral replacement using biodegradable polymer scaffolds seeded with urothelial and smooth muscle cells. *J Urol* 153 (suppl):4

Yoo JJ, Satar N, Atala A (1996): Bladder augmentation using biodegradable polymer scaffolds seeded with urothelial and smooth muscle cells. *J Urol* 155(suppl):5

9

TISSUE ENGINEERING AND THE VASCULAR SYSTEM

ROBERT M. NEREM, LINDA G. BRADDON, DROR SELIKTAR, AND THIERRY ZIEGLER

INTRODUCTION

For tissue engineering, i.e., the development of constructs composed of living cells and natural biological materials, the opportunities presented by the cardiovascular system are both important and challenging. Most of the effort to date in this area has been focused on blood vessel substitutes. This focus is motivated by the one-half million coronary artery bypass surgeries performed in the United States each year, with this figure having more than doubled in the last decade. Using native vessels, graft patency is 50%–70% after 10 years, with an observed acceleration of graft closure with time (The VA Coronary Artery Bypass Surgery Cooperative Study Group, 1992). Individuals who have had previous bypass surgery may not have additional native vessels available. Moreover, there is a significant percentage of the U.S. population for whom acceptable native vessels are not available. Synthetic materials such as Dacron and expanded polytetrafluoroethylene (ePTFE) have even lower graft patency rates than native vessels, and, as a result, these are not viable alternatives, especially for small diameter grafts. Thus, tissue engineering seems to offer the best possibilities for the next generation of vascular grafts.

Early approaches to the tissue engineering of a substitute blood vessel employed a synthetic biomaterial seeded with a monolayer of endothelial cells (EC). The purpose of the endothelial lining was to provide a "natural" interface between the flowing blood and underlying synthetic graft material. Some progress has been made with this approach (Williams et al, 1994; Zilla

Synthetic Biodegradable Polymer Scaffolds
Anthony Atala and David Mooney, Editors
Robert Langer & Joseph P. Vacanti, Associate Editors
© 1997 Birkhäuser Boston

et al, 1986); however, for small diameter vascular grafts such as needed for coronary bypass surgery, the success achieved to date still leaves much to be desired.

In the last decade a different tissue engineering approach has evolved. The goal of this has been the development of a hybrid blood vessel substitute involving a co-culture of EC and vascular smooth muscle cells (SMC), and it is these efforts that are the focus of this chapter. Considerable progress has been made in recent years, and the co-culture of EC and SMC using collagen gels will be discussed below. Although we are still a long way from clinical application, the use of such an approach in hybrid tubular constructs will be described. At least in some cases, a biomaterial of some type is employed, and the possible role of such synthetic biomaterials will be discussed.

Critical to the success of any cardiovascular implant is a recognition of the importance of the mechanical environment in which such a substitute must function. This mechanical environment is imposed by the hemodynamics of the vascular system and plays an important role in the regulation of the biology and pathobiology of blood vessels. For tissue-engineered substitutes, there is no mechanical environment more severe than that provided by the cardiovascular system, and in the next section both the characteristics of this environment and its effect on vascular biology will be discussed.

HEMODYNAMICS AND VASCULAR BIOLOGY/PATHOBIOLOGY

An important component of the environment in which vascular cells reside is the mechanical stresses resulting from the hemodynamics of the vascular system (Nerem, 1992). As blood flows through an artery, there are both transmural pressure acting normal to the vessel wall and a viscous shear stress acting in the tangential direction. EC at the interface with the flowing blood "sees" pressure and the shear stress directly, but also "rides" on a basement membrane which is being cyclically stretched as the pressure pulses. SMC, on the other hand, whether within the media or a thickened intima, are exposed to the stresses within the vessel wall, with the circumferential hoop stress being the major component of stress.

Although the magnitude of the shear stress is only a fraction of that of the pressure, there is now considerable evidence indicating an important role for shear stress in regulating endothelial biology. Much of what we know is from cell culture studies where endothelial monolayers have been exposed to well-defined flow conditions. From such experiments, we know that there are a variety of effects of flow and the associated shear stress on cultured endothelial cells. One of these is cell shape and orientation in which, as can be seen in Figure 1, EC elongate in response to flow and orient their major

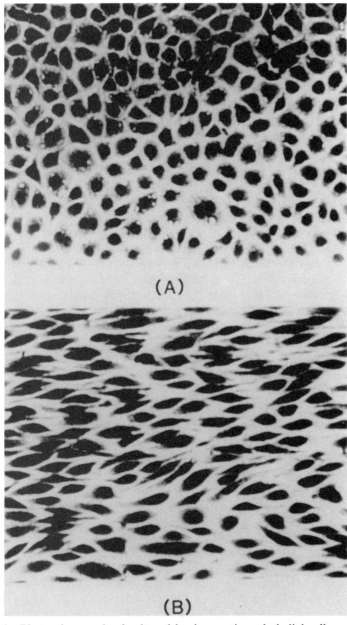

Figure 1. Photomicrograph of cultured bovine aortic endothelial cells grown on Thermanox under control conditions (A) and under a shear stress of 85 dynes/cm^2 for 24 hours (B); flow is from left to right. (Reprinted with permission from Levesque and Nerem, 1985).

axis with the direction of flow (Dewey et al, 1981; Eskin et al, 1984; Levesque and Nerem, 1985). This is in contrast to EC in static culture where cell morphology is characterized by a cobblestone appearance. The elongation response to flow exhibited by EC in culture is also observed *in vivo* (Levesque et al, 1986).

As dramatic as this elongation response is, there are other effects which are more important biologically. One of these is the decrease in cell proliferation, with the higher the shear stress, the lower the rate of cell growth. This is due to an inhibition of entry into S-phase (Levesque et al, 1990) and is not the result of cells detaching from the substrate. This thus indicates an effect on cell cycle. There also is an influence of flow on the synthesis and secretion of various biologically active molecules. This latter effect extends to the gene expression level, in some cases there being an upregulation, in others a downregulation, and in still others yet no effect at all. An example is the influence of flow and the associated shear stress on the release of nitric oxide (NO), a potent vasodilator but also a molecule possibly linked to the nature of the vessel wall's oxidative environment. In cell culture studies of the acute response to the onset of flow, there is a dramatic increase in NO release (Taylor et al, 1991). There also is a chronic increase which appears to be the result of an upregulation in nitric oxide synthase (NOS) which is required for the conversion of L-arginine to NO (Uematsu et al, 1995). PGI_2, another important vasodilator, is also increased by flow (Frangos et al, 1985), but for endothelin-1, a vasoconstrictor, the effect of flow is more complicated. Whereas low shear stress (5 dynes/cm^2) increases ET-1 secretion as well mRNA expression (Morita et al, 1993), high shear stress (25 cynes/cm^2) decreases secretion and mRNA expression (Sharefkin et al, 1991). There also seems to be a threshold level of shear stress which induces a change in the trend of secretion (Kuchan and Frangos, 1993). Moreover, the process of shear stress induced mRNA changes seems to occur via a disruption of the cytoskeletal structure (Morita et al, 1993). The influence of flow on the regulation of vasoactive molecules thus is a very differential one, and this is true in general in regard to the effect of flow on biologically active molecules.

Cyclic stretch also influences endothelial biology (Nerem, 1993; Nerem and Girard, 1990). This includes cell shape and orientation, cytoskeletal structure, cell proliferation, and the synthesis and secretion of various biologically active molecules. In some cases the influence of cyclic stretch is similar to that of flow. For example, in a recent study using elastic tubes, it was shown that shear stress and cyclic stretch act synergistically to enhance EC elongation and alignment with the tube long axis (Zhao et al, 1995). In other cases, however, the effect of cyclic stretch may produce a different, even opposite effect to that of flow. Thus, the regulation of endothelial structure and function by the hemodynamically-imposed mechanical environment is a very complex one.

Unlike shear stress and cyclic stretch, pressure appears to have no immediate effect on cell morphology. However, a week-long exposure to low levels of pressure around 15 cm of water was observed to induce cell elongation with no particular orientation, as well as an increase in cell proliferation (Acevedo et al, 1993). Pressure also acts on the secretion of vasoactive substances. ET-1 secretion is upregulated in a dose-dependent manner by a 4 to 8 hour exposure to pressures of 40 to 160 mm Hg (Hishikawa et al, 1995).

The influence of cyclic stretch on vascular SMC has also been studied in cell culture. Here there are equally dramatic effects, with cyclic stretch influencing cell orientation and a variety of other functional indicators. The latter include actin reorganization (Dartsch and Hammerle, 1986), increased collagen and total protein synthesis (Leung et al, 1975; Sumpio et al, 1988), and either unchanged or decreased cell proliferation, there being some discrepancy in the results reported in the literature. Here again the effects of cyclic stretch extend to the gene expression level. An example of this comes from recent studies conducted at Georgia Tech in which mRNA for the IGF-1 receptor was downregulated, while mRNA for MCP-1, monocyte chemotactic protein 1, was upregulated (Schnetzer et al, 1995). Finally, there is an indication that cyclic stretch may modulate the phenotype of SMC in culture (Birukov et al, 1995; Shirinsky et al, 1995).

It thus is clear that the mechanical environment of vascular cells alters their biological function. This is not only important to various disease processes, e.g., atherosclerosis, but also to the patency of any surgically-implanted blood vessel substitute.

ENDOTHELIAL CELL-SMOOTH MUSCLE CELL CO-CULTURE

Starting in the late 1970s, an approach began to evolve which employed a co-culture of EC and smooth muscle cells (SMC), together with appropriate extracellular matrix (ECM) components (Jones, 1979, Van Buul-Wortelboer et al, 1986). As promising as this approach is, to date it still is in the stage of basic research and development. The main motivation for this approach is that if one is to mimic nature, one presumably would need the same cell types as nature. In this case EC are there to provide a natural, dynamic, nonthrombogenic interface with flowing blood and to serve as a flow sensor. On the other hand, the SMC provide for vasoactivity and, together with the collagen and elastin, for the mechanical properties of the vessel.

In our own laboratory the motivation for moving to this type of model has been the goal of making cell culture studies of vascular biology more physiologic. As described in the previous section, initially our focus in this overall effort was on the hemodynamic environment, and our use of the parallel-plate flow chamber in which vascular EC could be exposed to well-defined conditions of flow and the associated shear stress represented a first step in this direction. A part of this approach also was to study effects of wall tension. This was done by growing either EC and SMC on a compliant membrane which could be subjected to a uniaxial cyclic stretch. Results of both types of studies, both from our laboratory as well as those of others, were briefly summarized in the last section.

As important as both these findings and the effects of physical force are, it is clear from ongoing research that there is more to the picture. Not only are EC and SMC influenced by their mechanical environment, but they also are affected by communication with each other (Hajjar et al, 1987; Herman, 1990). Thus, any successful model for the study of vascular biology needs to allow for interaction between these two cell types, which within the blood vessel wall are neighbors. Understanding the interaction of these neighboring cells also is required if one is to tissue-engineer a blood vessel substitute.

EC and SMC can interact by both humoral and direct contact mechanisms. In regard to the former, it has been shown that EC secrete both inhibitors and stimulants of SMC growth. When grown in co-culture with SMC, however, EC release more growth inhibitory activity (Xu et al, 1990). One of these inhibitory substances has been found to be a heparin-related glycosaminoglycan with a molecular weight of 10,000–15,000 (Castellot et al, 1981). Transforming growth factor β (TGF-β is another important molecule due to its multipotency (Schwartz and Liaw, 1993). TGF-β has been found to be the factor inhibiting EC growth when EC and SMC (or pericytes) are cultured in a culture system allowing direct contact between the two cell lines (Orlidge and D'Amore, 1987). The release of endothelin, a potent vasoconstrictor peptide and a SMC mitogen, by EC was shown by Stewart et al (1990) to be inhibited by co-culture with SMC. A final example is PDGF, which is known to stimulate different cell behavior *in vitro* than *in vivo*.

To study the influence of neighboring smooth muscle cells on EC growth in the presence of flow, we have developed in our laboratory a co-culture model of the arterial wall which is illustrated in Figure 2 (Ziegler et al, 1995a,b). In this co-culture model, porcine aortic SMC are seeded with soluble collagen I, and the collagen is allowed to polymerize in order to obtain a three-dimensional matrix with SMC. The cells bind to the collagen fibers and contracts the gel, with the final area depending on many factors including the number of SMC seeded. Once the collagen lattice has fully contracted, porcine aortic EC are plated at a very high cell density, with the cells rapidly becoming confluent. Extracellular matrix proteins are sometimes layered on top of the gel, underneath the EC. SMC grown in a collagen gel are much more elongated than cells grown on plastic and appear to have a phenotype close to contractile. Experiments have shown that the SMC grow very slowly in the gel as opposed to cells grown on a dish coated with

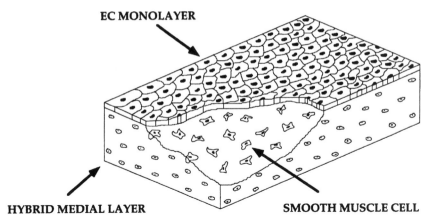

EC MONOLAYER

HYBRID MEDIAL LAYER

SMOOTH MUSCLE CELL

Figure 2. Illustration of co-culture model of a blood vessel wall. The cobblestone EC morphology is characteristic of static culture, whereas the SMC within the collagen gel are depicted as having a random orientation typical of an unstressed gel.

collagen. EC grown on plastic have a cobblestone shape at confluence; however, when the cells are seeded on top of the SMC-collagen gel construct, they are much more elongated. The complete coverage of the EC and their elongation has been confirmed by scanning electron microscopy (Ziegler et al, 1995b). No SMC were seen on top of the endothelium up to 10 days after EC seeding.

Such EC-SMC collagen constructs have been exposed to the onset of flow, with experiments to date focusing on morphology and on cell proliferation. Using shear stresses of 10 and 30 dynes/cm^2, both EC elongation and alignment have been measured. Since, as noted above, the EC are elongated even under static culture conditions, there is little further elongation with flow. However, there is a definite alignment of the EC monolayer in the direction of flow, with this occurring more rapidly at 30 dynes/cm^2 than at 10 dynes/cm^2.

The results on EC proliferation have been of particular interest to us (Ziegler et al, 1995a). A typical EC monolayer in static culture, even when confluent, exhibits a cell turnover rate far in excess of that observed in animal studies. Whereas *in vivo* the endothelium may be characterized as being quiescent, this is not true for EC monolayers in culture. Also, as noted in the previous section, an EC monolayer, cultured on tissue culture plastic and exposed to the onset of a steady flow, exhibits a decrease in cell proliferation. The question then for this new co-culture model was what would be the proliferative activity of an EC monolayer seeded on top of a collagen-SMC gel? As presented in Table 1, the results indicate that, even under static culture conditions, the endothelium exhibits a low cell turnover rate. As previously reported (Ziegler et al, 1995a), the growth

Table 1. Effect of culture conditions on the thymidine uptake by porcine EC and SMC

Culture Condition	^{3}H-Thymidine Incorporation cpm/1000 cells (mean $\pm$ sem)
Confluent EC on plastic 96 hr after seeding at 3.2×10^4 cells/cm^2	160 ± 37
EC exposed to 30 dyn/cm^2 for 24 hr	109 ± 23
Confluent EC on collagen gels 96 hr after seeding at 3.2×10^4 cells/cm^2	42 ± 22
Co-culture of EC and SMC	35.5 ± 10.9
Co-culture exposed to 30 dyn/cm^2 for 2 hr	25.3 ± 9.4

rate of EC in static co-culture was found to be 35 cpm/1000 cells. This was at best only slightly lower than that for a collagen gel without SMC, and the measured EC growth rate for both of these cases represented a factor of five decrease compared with PAEC on plastic. Furthermore, this corresponds to the presence of approximately 1% dividing cells and is a number very similar to that found in the endothelium of blood vessels (Freudenberg, 1985), indicating that ECM proteins in a fibrous form may be a prerequisite for the conservation of a differentiated, quiescent endothelium. With such low growth rates, any effect of flow on the growth rate of co-cultured EC was difficult to observe, although there did appear to be a further slight decrease.

These initial studies demonstrate that it is possible to reconstitute a model of the vascular wall in cell culture where the EC make a monolayer which covers the surface of the gel and within which reside SMC in a quiescent condition. Most importantly, the endothelium appears to be in a quiescent state. Even so, there is much to be done, particularly in regard to the vasoactivity and the thrombogenicity of such a construct. Furthermore, there is a need to increase the SMC density in such collagen gels to a level comparable to normal tissue.

HYBRID TUBULAR CONSTRUCTS

A next step for our laboratory has been the development of an EC-SMC co-culture, collagen-based tubular construct. Such a configuration may well require a synthetic biomaterial, if not as a scaffold, then perhaps to provide structural integrity, and the general role of biomaterials will be discussed in the next section. In this section, however, we will review progress to date in developing collagen-based tubular constructs.

The credit for the development of the first collagen-cell construct must be given to Bell et al (1979) who created a tissuelike substance from rat tail collagen and fibroblasts. The addition of the fibroblasts caused a contraction of the collagen lattice and opened the door for future research in tissue replacements. For this work with fibroblasts in a collagen gel, the application of immediate interest was the development of this novel substrate as a skin equivalent. However, in the mid-1980s Weinberg and Bell (1985, 1986) made a significant contribution to blood vessel substitutes by the development of a construct made with vascular cells and collagen. In this the intimal region of the vascular equivalent was made with collagen and bovine smooth muscle cells. The collagen lattice was allowed to contract for a week, at which point a Dacron mesh sleeve was slipped over the structure in order to add structural integrity. Next, a similar collagen gel, this time with fi-broblasts, was molded to the outer surface of the substrate, thereby forming a vascular equivalent with two distinct regions mimicking the media and the adventia. The inner surface was then seeded with endothelial cells, producing a monolayer which reached approximately 92% confluency.

Bursting strength tests were conducted on specimens by exposing them to increased lumenal pressure. The results indicate that the collagen with cells alone was not strong enough to withstand physiological pressures; however, layering of multiple collagen sections and reinforcement with Dacron meshes produced a much stronger composite material capable of withstanding physiological pressures. The burst strength was a function of the degree of collagen contraction and the number of layers alternated with reinforcing meshes.

The work of Weinberg and Bell was expanded by Auger and his coworkers using human cells and collagen to construct a triple-layer tubular vascular construct without the reinforcement of a Dacron sleeve (L'Heureux et al, 1993). By culturing their medial equivalent on a mandrel, they observed alignment of the smooth muscle cells in a circumferential manner with the occurrence of intercellular junctions. The degree of orientation of the smooth muscle cells was also observed to be a function of their radial locations and their time in culture. SMC that were adjacent to the casting mandrel exhibited a higher degree of circumferential alignment than those cells on the outer radial periphery. Likewise, a higher percentage of the SMC showed radial alignment at longer times in culture.

When this tubular construct was seeded with human umbilical vein EC, the cells became confluent on the lumenal surface of the construct. However, a major difference between these tissue engineered vascular equivalents and the native artery was the density of the smooth muscle cells. In these constructs, the SMC density was approximately an order of magnitude lower than that seen *in vivo*. When SMC density was increased in the collagen lattice, overgrowth of the endothelium occurred by both migration and proliferation of the medial smooth muscle cells.

Matsuda and coworkers have used two approaches to the development of a hybrid tubular construct. In one approach they use a compliant, porous polyurethane graft to reinforce a hybrid medial layer comprised of SMCs, type I collagen, and dermatan sulfate (Miwa et al, 1993). It has been shown that dermatan sulfate enhances adhesion and growth of endothelial cells and reduces the adhesion of platelets *in vitro*. A solution of collagen, dermatan sulfate, and cells is applied onto the lumenal surface of the polyurethane graft by repeated cycles of coating and thermal gelation. The hybrid tissue is then seeded with an EC monolayer on the lumenal surface. In the second approach, a hybrid medial layer is prepared with a mixture of SMCs and type I collagen molded between a mandrel and a glass tube (Hirai et al, 1994). After incubation and subsequent removal of the mandrel, the hybrid medial tissue is seeded with EC onto the lumenal surface to produce a hierarchically structured hybrid vascular tissue. The size and mechanical properties of the construct depend on SMC seeding density, collagen concentration, incubation time, and geometric parameters. The hybrid vascular tissue developed by the second method resembles the native structure more precisely but lacks the mechanical strength to withstand physiological pressures.

Tissue-engineered tubular constructs developed under the first approach yield a more suitable prothsesis for *in vivo* studies. The results from such studies reveal several key elements about the reorganization of tissue engineered vascular grafts subjected to physiological conditions. On the lumenal surface of the graft, an endothelial cell monolayer aligns in the direction of the blood flow. The medial layer increases in thickness and SMCs redistribute and become circumferentially oriented. The resulting construct resembles the native arterial structure more closely than its unmodified counterpart. To date, no *in vivo* studies have been performed on the hybrid vascular tissue developed by the second approach.

A critical question is the phenotype of vascular SMC within a blood vessel substitute. It is well known that when SMC are put into culture, they undergo a modulation from their normal contractile phenotype *in vivo* to a synthetic phenotype. For a blood vessel substitute that incorporates SMC, it clearly would be advantageous to induce a phenotypic modulation back to the contractile state. There are many factors that might contribute to such a modulation. This includes the presence of the neighboring EC (Campbell and Campbell, 1986), the three-dimensional organization of the vessel (Kanda et al, 1993a,b), and the influence of the mechanical environment (Shirinsky et al, 1995). Even with these insights, however, there still is much to be done if we are to revert SMC back to a contractile phenotype.

Important to the mechanical properties of an SMC-collagen construct is the orientation of the SMC. Tranquillo et al (1996) have shown that it is possible to orient the SMC by applying a strong magnetic field during fibril formation. This not only increases the stiffness of the construct compared to control specimens but also results in reduced creep in the circumferential direction.

It should be noted that blood vessels exhibit very complicated material property responses such as nonlinear stress-strain characteristics and also stress relaxation, creep, and hysteresis, which serve to classify an artery as a viscoelastic material (Fung, 1981). The ideal vascular graft would not only be strong enough to withstand physiologically relevant pulsatile pressures but would also match the compliance values of a native artery.

The proper characterization of the mechanical properties of collagen-based cellular substitutes has in fact been somewhat neglected. In preliminary studies conducted in our laboratory to determine the mechanical properties of a collagen-SMC lattice, a nonlinear, plasticlike material response was exhibited (Greer et al, 1994). The critical breaking stress of the gel was found to be largely dependent on cell seeding density, and the stress-strain characteristics depended on strain rate. For example, for a strain velocity of 2.3 mm/sec, gels seeded with an innoculum of 1 million cells had a critical breaking stress of 23,100 N/m^2, whereas for gels seeded at 2 million cells the critical breaking stress was 33,000 N/m^2. The results of this study also indicate that such gels show certain viscoelastic properties, such as stress relaxation, which are characteristic of biological tissue. The strength of the collagen gel also was found to be a function of additives to the collagen-SMC construct. For example, the addition of the glycosaminoglycan, chitosan, inhibits gel contraction and alters the strength of the material.

It is clear that collagen gels, whether cell-seeded or not, do not match the strength of a native blood vessel or exhibit the elasticity present in native vessels. The result of a lack of elasticity is a plasticlike response and the inability of the gel to recover from the application of strain. Therefore, there is a need to add elasticity to these tubular constructs. This can be achieved either by incorporating elastin itself or a synthetic elastic material. The former possibly could be achieved by inducing the SMC, or some other cell type, to synthesize and secrete elastin, while the latter might involve the use of a synthetic biomaterial as will be discussed in the next section. In addition to there being elastin present, there needs to be an organization into elastic laminae. van der Lei et al (1986) has shown that compliance induces the regeneration of elastic laminae within the vessel wall.

ROLE OF SYNTHETIC BIOMATERIALS

The use of synthetic biomaterials as vascular substitutes was initiated in the early 1950s with the development of plastics and other polymeric substances (Hastings, 1992; Hess, 1985; Yeager and Callow, 1988). The use of nonsolid, porous, woven, braided, or knitted fabrics as vascular protheses followed soon after. The first such biomaterial to be used clinically as a vascular graft was a woven fabric prosthesis made of Vinyon N. Vascular prostheses have also been made from nylon, Teflon, Orlon, Dacron, polyethylene, and polyurethane. Nylon was eliminated due to its rapid degeneration *in vivo*. Of the remaining materials, a comparison of the mechanical properties distinguished Dacron as the most suitable for a blood vessel substitute. In the mid-1970s ePTFE was introduced as another suitable material for vascular prosthesis. In the early 1980s a novel approach to vascular grafts was introduced with the development of bioabsorbable materials which degrade over time while cellular components are being produced to take up the structural role.

Today, tissue-engineered blood vessel substitutes incorporate many of the same synthetic biomaterials used for artificial vascular grafts. Dacron, polyurethane, and ePTFE grafts have all been used to reinforce hybrid vascular tissue. A critical problem with current vascular grafts is the adhesion of platelets to the graft surface. To prevent this, synthetic grafts have been coated with biological substances, e.g. collagen, albumin, heparin, fibronectin, and dermatan. The purpose has been to reduce the adhesion of platelets, to prevent activation of the clotting system, and to enhance endothelial cell growth. Bioabsorbable graft materials such as polyglycolic acid and poly-L-lactic acid are used as a transitional element providing structure for the developing tissue. Another tissue-engineering-based blood vessel substitute employs a photocrosslinkable, water soluble polymer and SMCs to create a synthetic polymer that mimics the media of the vascular wall (Miwa et al, 1991).

A focus of the efforts by Hubbell and West (1995) has been on the development of hydrogels for use as a coating on an injured arterial surface or a vascular graft. The intent is to provide a barrier between the flowing blood and the underlying arterial or graft surface. In the former case, experiments in which an injured arterial wall was covered by a hydrogel resulted in the reduction of platelet adhesion by as much as 90%. There also was a decrease in the SMC proliferation associated with the healing response. Such a hydrogel coating also could be engineered to enhance EC attachment and spreading (Hubbell et al, 1991).

As noted previously, there are different roles that a synthetic biomaterial can play in a tissue-engineered cardiovascular substitute. Two key ones are (1) a scaffold to provide for the three-dimensional architecture, and (2) a structural material to impart the appropriate mechanical properties to the construct. The need of a three-dimensional structure is no less for a blood vessel substitute than for other tissue-engineered constructs, and much has already been said in earlier chapters about synthetic biodegradable polymer scaffolds, the focus of this book. If one views the construct as a transitional one, then over a period of time the host's own cells will take over from those initially seeded into the construct. In this case, a synthetic biodegradable scaffold not only can provide a transitional structure, but it also can deliver biologically active molecules, e.g., growth factors or chemotactic proteins, to enhance certain cellular processes. One may wish to enhance EC recruitment to promote endothelialization of the substitute with the host's own EC. Similarly, one might want to promote SMC recruitment and proliferation. One concept that has been proposed is the use of a biomaterial scaffold seeded with fibroblasts, as opposed to SMC. The choice of the fibroblast cell is based on its availability from human foreskin and the fact that fibroblasts have only a low level expression of the antigen associated with immune rejection, and thus autologous cells can be employed. It should be noted that the biomaterial scaffold could be of the biodegradable polymer type, but it could also be decellularized tissue.

In addition to the use of a biomaterial as a transitional scaffold, it also could be used as a structural material. The previously noted work by Matsuda and his coworkers includes this type of application, and in this case the biomaterial is not biodegradable, thus not playing a transitional role but rather one of a more permanent nature (Miwa et al, 1993). It should be emphasized that the mechanical properties of a blood vessel substitute may be very important as there are those who believe that there should be impedance matching between the implanted substitute and the native system. This means that it is not sufficient for a blood vessel substitute to simply be able to withstand the hemodynamic loading imposed. Such a substitute must have mechanical properties that mimic those of a normal blood vessel.

In this regard the studies of Urry and his coworkers are of interest (Urry et al, 1995). In their more recent work, they are producing bioelastic materials based on elastomeric and related polypeptides through the use of recombinant DNA technology. In particular, the bioelastic material X^{20}-poly (GVGIP) made using *Escherichia coli* as the host cell exhibits a Young's modulus in the range of 4 to 6×10^6 dynes/cm^2. This is very similar to that of a normal blood vessel.

Although there are many issues that have not been addressed here, there is a final comment to be made relative to the use of a synthetic biomaterial for the tissue engineering of cardiovascular substitutes. This is in regard to the interaction between the biomaterial and the cells employed. In a tissue-engineered construct, a biomaterial serves as a substrate for the adhering cells and as such is part of the local cellular environment. Cell function will be determined not only by the biochemical environment and the imposed hemodynamic loading, but also by the substrate to which the cells are anchored. An excellent example of this is provided in Figure 3. This picture shows an endothelial monolayer cultured on a vicryl biomaterial which has been used by Advanced Tissue Sciences (La Jolla, CA) in many of their tissue-engineered products. As may be seen, for the static culture conditions of this study the morphology is considerably different than the cobblestone pattern normally observed in which there is a random cellular orientation

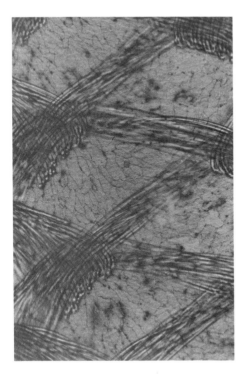

Figure 3. Photomicrograph of endothelial cells cultured under static conditions on a vicryl substrate provided courtesy of Advanced Tissue Sciences, Inc. (LaJolla, CA). The pattern of the vicryl substrate has a significant influence on endothelial morphology which is clearly different from that shown in Figure 1A.

(see Figure 1). With the vicryl substrate, the pattern of the biomaterial clearly influences EC shape and orientation. Once such an EC-substrate construct is exposed to flow, although there is a tendency to orient as is normally seen, the degree of alignment is influenced, with the cell orientation being the product of the direction of flow and the associated shear stress and the pattern of the underlying biomaterial. It must be assumed that this influence of the underlying substrate is not just on morphology, but also and even more importantly on cell function, e.g., the synthesizing and secretion of vasoactive substances and growth factors and the expression of cell adhesion molecules.

DIRECTIONS FOR THE FUTURE

It is clear from the preceding that there is still much to be done if we are to successfully tissue-engineer vascular substitutes for clinical application. In the short term, those companies in industry that started their tissue engineering research focusing on skin equivalents appear to have the edge. Two competitors in this field are Advanced Tissue Sciences (La Jolla, CA) and

Organogenesis (Canton, MA). Both companies are attempting to tissue-engi-neer a reliable small diameter vascular graft. Organogenesis is using a pro-prietary material called Dense Fibrillar Collagen (DFC) which can be formed into cables, braids, knits, weaves and other various substrates. This DFC mate-rial is used in conjunction with a small intestinal submucosa which is stated to be nonimmunogenic, suturable, and exhibiting acceptable strength charac-teristics. Advanced Tissue Sciences is using a biodegradable polymer base with human dermal fibroblasts. The biodegradable polymer supplies a tempo-rary scaffold for the living cells which can absorb stress until the cells produce the matrix products necessary to produce a stable tissue equivalent.

What about the long term? What are the research directions required for the future? An excellent start on this is provided by the Tissue Engineering Working Group convened by the National Heart, Lung, and Blood Institute (NHLBI) in July 1995 during the Gordon Research Conference on Biocom-patibility and Biomaterials. As part of the Working Group's report (Ander-son et al, 1995), the following general principles were expressed: (1) cross-disciplinary research and a team approach, often embraced but rarely implemented, are particularly essential for progress in this field; (2) ap-proaches and concepts require an integration of several different disciplines, e.g., cardiovascular system biology, physiology, anatomy, microanatomy, pa-thology, biochemistry, pharmacology, developmental biology, and materials science; (3) detailed consideration of the cell and molecular biology of cardiovascular cells should be given; (4) factors influencing success or failure of tissue-engineered cardiovascular implants need to be identified; and (5) device development requires utilization of device design criteria at the earliest stages of research.

The Working Group, in discussing a variety of research topics, organized its recommendations as follows. Under the category of new biomaterials designed for tissue-engineered cardiovascular implants, the following priori-ties were identified: (1) biomimetic materials endowed with cell or cell-based signals; synthetic extracellular matrix for enhanced cell interaction, cell polarization, or remodeling; temporal and/or spatial delivery of bioac-tive agents over short and long time periods; (2) biomaterials whose chemi-cal, physical or mechanical properties, structure, or form permits active tissue integration of desirable cell types and tissue components; (3) biointer-active and environmentally responsive materials with controllable biodegra-dation, cell adhesion, cell activation, and biocompatibility; and (4) computer aided design for material macro-, micro-, and ultrastructure to facilitate tissue-engineered implants.

The understanding of biological signals and signaling mechanisms in car-diovascular implants also was identified as important. Here the priorities are: (1) mechanisms by which synthetic extracellular matrices induce cell signaling and subsequent cellular responses; (2) integrin and other receptor signal transduction by surfaces, tethered ligands, bioactive agents, and com-

180 Nerem et al.

bination of these; (3) signaling mechanisms that influence or determine biomaterial encapsulation versus integration; (4) the role of apoptosis, ischemia and immune mechanisms in the failure of tissue-engineered cardiovascular implants; and (5) the role, mechanisms, and regulation of transduced mechanical forces in healing and tissue integration.

A third area of importance is both the normal and directed healing mechanisms in cardiovascular implants. In this category the priorities are: (1) the development and evaluation of resorbable templates that direct tissue formation; (2) the determination of the role of nitric oxide, cytokines, growth factors, and other mediators in healing processes; (3) characteristics of materials designed to modulate healing and angiogenesis; (4) mechanisms of communication among cardiovascular cell types as well as among cells of the same type in relation to the production and maintenance of extracellular matrix on cardiovascular implants; and (5) the induction of matrix development and turnover by cardiovascular implants.

A fourth area is the delivery and phenotypic expression of cells in and on cardiovascular implants. Specific topics here include: (1) novel methods for the delivery of genes to vascular cells in or on cardiovascular implants or tissue sites with controlled transcription and targeted cell responses; (2) enhanced understanding of the factors controlling vascular cell phenotypic expression in implants when tested *in vitro* compared to *in vivo;* (3) optimization of methods for inducing the expression of native or introduced genes in vascular cells in tissue engineered implants; (4) development of media, synthetic gels, or vehicles for optimized behavior of vascular cells in tissue engineered implants; and (5) development of methods to monitor the metabolism, desired response, or other parameters related to viability and function of cardiovascular cells in tissue engineered implants.

The working group also identified gender-related research issues. One of these is the effects of estrogen on functions of cardiovascular cells in tissue engineered implants. They also made several recommendations relative to the role of the cardiovascular system in bone resorption and healing in temporomandibular joint (TMJ) reconstruction and in TMJ prosthesis failure. These, however, do not have specific application to vascular substitutes and thus will not be discussed here.

Many of the noted topics above have been briefly discussed here; however, there are several aspects that we would like to emphasize. First is the necessity of controlling interactions at the interface between a synthetic biomaterial, whatever its purpose, and neighboring cells. This includes both the cells seeded in a construct as well as the host's own cells. With the proper design of the surface of a biomaterial, or through the controlled release of biologically active molecules, it will be possible to engineer cell adhesion, motility, and/or proliferation.

Second is the role of the vascular mechanical environment. This was included in the discussion of biological signals and signaling mechanisms, and as indicated, there is a role for mechanical, physical forces in the regulation of healing and tissue integration. Although much has been learned over the past decade about the influence of flow and the associated shear stress and about the effect of cyclic stretch, there is still much that is unknown. What about synergistic effects between shear stress and cyclic stretch? What if the endothelial monolayer is adhering to a moving surface, which, at least in the case of a heart valve substitute, could be a major influence? How important is the mechanical environment of a blood vessel in determining the phenotype of SMC? What is the nature of the interaction between a cell's hemodynamic environment and the biomaterial to which it is interacting? Is it possible that certain mechanical environments are in fact proinflammatory?

A final area to be emphasized is the genetic manipulation of cells to be seeded in a construct so as to enhance certain cell function characteristics. For example, one may wish to enhance EC motility and proliferation so as to accelerate the formation of a confluent monolayer, or alternatively it may be desirable to enhance cell adhesion. Another example is to engineer a cell type to make the connective tissue components, e.g., elastin, necessary in a three-dimensional vessel structure.

These and many other questions still need to be answered before we will be able to successfully tissue-engineer substitutes for the vascular system.

Concluding Discussion

Even though it is clear that much progress has been made in just the last decade, there is still much to be done. The result is that we are still approximately a decade away from the day when tissue-engineered substitute blood vessels are used routinely as vascular grafts clinically in coronary bypass surgery. This is true even of those approaches that have already moved into the commercial sector. For those still in the university and medical center research laboratories, it will be even longer before their application clinically is realized.

As the field of tissue engineering continues to evolve, an additional vascular need has presented itself, the tissue engineering of prosthetic heart valves. Nearly 100,000 prosthetic heart valves are implanted each year in the United States, with 70% of these being mechanical valves and the remainder bioprosthetic valves. However, none of these exhibit opening and closing dynamics which mimic the native valve. There is no mechanical environment more severe than that associated with a heart valve. Not only does one have the effects of flow and the associated shear stress coupled with stresses with the valve leaflets themselves, but there is also the dynamics of opening and closing, which occur at a frequency of approximately 1 Hz. Although the medical implant industry now views the next generation of prosthetic heart valves as ones that will be tissue-engineered, to tissue-engineer a heart valve substitute will be a formidable challenge, and in the following chapter this topic is addressed. Although several groups already are at work, it will take even longer to achieve success here than in the case of blood vessel substitutes.

There thus are a variety of challenges confronting those focusing on tissue engineering and the vascular system. This is true both for blood vessels and heart valves. To achieve success will require interdisciplinary as well as multidisciplinary efforts. Such teams will need to include specialists in bioengineering, biomaterials, cell biology, immunology, and vascular surgery. In bringing vascular products to market, the efforts may need to include not only industrial teams but also the active participation of academic researchers. Whatever the case, there is still much to be done, both to bring the first generation of products to market and to further the basic understanding necessary to provide a foundation for the improvements that will characterize the next generation of tissue-engineered vascular substitutes.

Acknowledgments   This work was supported by National Science Foundation grants BCS-911761 and BES-9412010. The authors thank the other members of our laboratory and our collaborators for the many discussions that have led to the ideas reflected in this paper and for their participation in the study of this subject.

References

Acevedo AD, Bowser SS, Gerritsen ME, Bizios R (1993): Morphological and proliferative responses of endothelial cells to hydrostatic pressure: Role of fibroblast growth factor. *J Cell Physiol* 157:603–614

Anderson (1995): Tissue engineering in cardiovascular disease: A report. *J Biomed Mat Res* 29:1473–1475

Bell E, Ivarsson B, Merrill C (1979): Production of a tissue-like structure by contraction of collagen lattices by human fibroblasts of different proliferative potential in vitro. *Proc Natl Acad Sci* 76:1274–1278

Birukov KG, Shirinsky VP, Stepanova OV, Tkachuk VA, Hahn AW, Resink TJ, Smirnov VN (1995): Stretch affects phenotype and proliferation of vascular smooth muscle cells. *Mol Cell Biochem* 144(2):131–139

Campbell JH, Campbell GR (1986): Endothelial cell influences on vascular smooth muscle phenotype. *Annu Rev Physiol* 48:295–306

Castellot JJ, Addonizio ML, Rosenberg R, Karnovsky MJ (1981): Cultured endothelial cells produce a heparin-like inhibitor of smooth muscle cell growth. *J Cell Biol* 90:372–379

Dartsch PC, Hammerlee, H (1986): Orientation of cultured arterial smooth muscle cells growing on cyclically stretched substrates. *Acta anat* 125:108–113

Dewey CF, Bussolari SR, Gimbrone MA Jr, Davies PF (1981): The dynamic response of vascular endothelial cells to fluid shear stress. *ASME J Biomech Eng* 103:177–181

Eskin SG, Ives CL, McIntire LV, Navarro LT (1984): Response of cultured endothelial cells to steady flow. *Microvasc Res* 28:87–94

Frangos JA, McIntire LV, Eskin SG, Ives CL (1985): Flow effects on prostacyclin production by cultured human endothelial cells. *Science* 227:1477–1479

Freudenberg, N. General properties of endothelial cells. In: Schettler G, Nerem RM, Sclunid-Schonbein H, Morl H, Diehm C, eds (1985): *Fluid Dynamics as a Localizing Factor for Atherosclerosis* Berlin: Springer-Verlag

Fung YC (1981): *Biomechanics: Mechanical Properties of Living Tissues.* New York: Springer-Verlag

Greer LS, Vito RP, Nerem RM (1994): Material property testing of a collagen-smooth muscle cell lattice for the construction of a bioartificial vascular graft. *ASME Adv Bioeng* 28:69–70

Hajjar DP, Marcus AJ, Hajjar KA (1987): Interactions of arterial cells: Studies on the mechanisms of endothelial cell modulation of cholesterol metabolism in co-cultured smooth muscle cells. *J Biol Chem* 262:6976–6981.

Hastings GW, ed. (1992): *Cardiovascular Biomaterials.* London: Springer-Verlag

Herman IM (1990): Endothelial cell matrices modulate smooth muscle cell growth, contractile phenotype and sensitivity to heparin. *Haemostasis* 20(suppl 1):166–177

Hess F (1985): History of (micro) vascular surgery and the development of small-caliber blood vessel prostheses (with some notes on patency rates and re-endothelialization). *Microsurgery* 6:59–69

Hirai J, Kanda K, Oka T, Matsuda T (1994): Highly oriented, tubular hybrid vascular tissue for a low pressure circulatory system. *ASAIO J* 40:M383–M388

Hishikawa K, Nakaki T, Suzuki H, Kato R, Saruta T (1995): Pressure enhances endothelin-1 release from cultured human endothelial cells. *Hypertension* 25:449–452

Hubbell JA, West JL (1995): Hydrogels for manipulating post-surgical healing. In: Nerem RM. *Proceedings of the 2nd International Conference on Cellular Engineering* La Jolla, CA: International Federation of Medical and Biological Engineering

Hubbell JA, Massia SP, Desai NP, Drumheller PD (1991): Endothelial cell-selective materials for tissue engineering in the vascular graft via a new receptor. *Biotechnology* 9:568–572

Jones PA (1979): Construction of an artificial blood vessel wall from cultured endothelial and smooth muscle cells. *Cell Biology* 76:1882–1886

Kanda K, Matsuda T, Miwa H, Oka T (1993a): Phenotypic modulation of smooth muscle cells in intim-media incorporated hybrid vascular prostheses. *ASAIO J* 39:M278–M282

Kanda K, Matsuda T, Oka T (1993b): In vitro reconstruction of hybrid vascular tissue hierarchic and oriented cell layers. *ASAIO J* 39:M561–M565

Kuchan MJ, Frangos JA (1993): Shear stress regulates endothelin-1 release via protein kinase C and cGMP in cultured endothelial cells. *Am J Physiol* 264 (33):H150–H156

Leung DYM, Glagov S, Mathews MB (1975): Cyclic stretching stimulates synthesis of matrix components by arterial smooth muscle cells *in vitro. Science* 191/4226:475–477

Levesque MJ, Nerem RM (1985): The elongation and orientation of cultured endothelial cells in response to shear stress. *ASME J Biomech Eng* 176:341–347

Levesque MJ, Liepsch D, Moravec S, Nerem RM. (1986): Correlation of endothelial cell shape and wall shear stress in a stenosed dog aorta. *Arteriosclerosis* 6:220–229

Levesque MJ, Nerem RM, Sprague EA (1990): Vascular endothelial cell proliferation in culture and the influence of flow. *Biomaterials* 11:702–707

L'Heureux N, Germain L, Labbe R, Auger FA (1993): In vitro construction of a human blood vessel from cultured vascular cells: A morphological study. *J Vasc Surgery* 17(3):499–509

Miwa H, Matsuda T, Kondo K, Tani N, Fukaya Y, Morimoto M, Iida F (1991): Development of a 3-D artificial extracellular matrix. Design concept and artificial vascular media. *ASAIO Trans* 37:M437–M438

Miwa H, Matsuda T, Iida F (1993): Development of a hierarchically structured hybrid vascular graft biomimicking natural arteries. *ASAIO J* 39:M273–M277

Morita T, Kurihara H, Maemura K, Yoshizumi M, Yazaki Y (1993): Disruption of cytoskeletal structures mediates shear stress-induced endothelin-1 gene expression in cultured aortic endothelial cells. *J Clin Invest* 92:1706–1712

Nerem RM (1992): Vascular fluid mechanics, the arterial wall, and atherosclerosis. *ASME J Biomech Eng* 114:274–282

Nerem RM (1993): Hemodynamics and the vascular endothelium. *ASME J Biomech Eng* 115:510–514

Nerem RM, Girard PR (1990): Hemodynamic influences on vascular endothelial biology. *Toxicol Pathol* 18:572–582

Orlidge A, D'Amore P (1987): Inhibition of capillary endothelial cell growth by pericytes and smooth muscle cells. *J Cell Biol* 105:1455–1462

Schnetzer KJ, Delafontaine P, Nerem RM (1995): Uniaxial cyclic stretch of rat aortic smooth muscle cells. *Ann Biomed Eng* 23(suppl 1):42

Schwartz SM, Liaw L (1993): Growth control and morphogenesis in the development and pathology of arteries. *J Cardiovasc Pharmacol* 21(suppl 1):31–49.

Sharefkin JB, Diamond SL, Eskin SG, Dieffenbach C, McIntire LV (1991): Fluid flow decreases endothelin mRNA levels and suppresses endothelin peptide release in human endothelial cells. *J Vasc Surg* 14:1

Shirinsky VP, Birukov KG, Stepanova OV, Tkachuk VA, Hahn AWA, Resink TJ (1995): Mechanical stimulation affects phenotype features of vascular smooth muscle cells. In: Woodford FP, Davignon eds. *Atherosclerosis X, Proceedings of the 10th International Symposium on Atherosclerosis.* Montreal, Canada: Elsevier

Stewart DJ, Langleben D, Cernacek P, Cianflone K (1990): Endothelin release is inhibited by coculture of endothelial cells with cells of vascular media. *Am J Physiol* 259:H1928–H1932

Sumpio BE, Banes AJ, Link WG, Johnson G (1988): Enhanced collagen production by smooth muscle cells during repetitive mechanical stretching. *Arch Surg* 123:1233–1236

Taylor WR, Harrison DG, Nerem RM, Peterson TE, Alexander RW (1991): Characterization of the release of endothelium-derived nitrogen oxides by shear stress. *FASEB J* 56:A1727

Tranquillo RT, Girton TS, Bromberek BA, Triebes TG, Mooradian DL (1996): Magnetically-oriented tissue-equivalent tubes: Application to a circumferentially-oriented media-equivalent. *Biomaterials:* in press

Uematsu M, Ohara Y, Navas JP, Nishida K, Murphy TJ, Alexander RW, Nerem RM,

Harrison DG (1996): Regulation of endothelial cell nitric oxide synthase mRNA expression by shear stress. *Am J Physiol Cell Physiol:* in press

Urry DW, Nicol A, McPherson DT, Xu J, Shewry PR, Harris CM, Parker TM, Gowda DG (1995): Properties, preparations and applications of bioelastic materials. In: *Handbook of Biomaterials and Applications.* New York: Marcel Dekker

The VA Coronary Artery Bypass Surgery Cooperative Study Group (1992): Eighteen-year follow-up in the veterans affairs cooperative study of coronary artery bypass surgery for stable angina. *Circulation* 86:121–130

Van Buul-Wortelboer MF, Brinkman HJM, Dingemans KP, DeGroot PG, van Aken WG, van Mourik JA (1986): Reconstruction of the vascular wall *in vitro:* A novel model to study interactions between endothelial and smooth muscle cells. *Exp Cell Res* 162:151–158

vander Lei B, Wildevuur CRH, Nieuwenhuis P (1986): Compliance and biodegradation of vascular grafts stimulate the regeneration of elastic laminae in neoarterial tissue: An experimental study in rats. *Surgery* 99:45–52

Weinberg CB, Bell E (1985): Regulation of proliferation of bovine aortic endothelial cells, smooth muscle cells and dermal fibroblasts in collagen lattices. *J Cell Physiol* 122:410–414

Weinberg CB, Bell E (1986): A blood vessel model constructed from collagen and cultured vascular cells. *Science* 231:397–399

Williams SK, Jarrell BE, Kleinert LB (1994): Endothelial cell transplantation onto porcine arteriovenous grafts evaluated using a canine model. *J Invest Surg* 7:503–517

Xu CB, Falke P, Stavenow L (1990): Interactions between cultured bovine arterial smooth muscle cells and endothelial cells: Studies on the release of growth inhibiting and growth stimulating factors. *Artery* 17(6):297–310

Yeager A, Callow AD (1988): New graft materials and current approaches to an acceptable small diameter vascular graft. *ASAIO Trans* 34:88–94

Zhao S, Suciu A, Ziegler T, Moore JE Jr., Burki E, Meister JJ, Brunner HR (1995): Synergistic effects of fluid shear stress and cyclic circumferential stretch on vascular endothelial cell morphology and cytoskeleton. *Arterioscler Thromb Vasc Biol* 15:1781–1786

Ziegler T, Alexander RW, Nerem RM (1995a): An endothelial cell-smooth muscle cell co-culture model for use in the investigation of flow effects on vascular biology. *Ann Biomed Eng* 23:216–225

Ziegler T, Robinson KA, Alexander RW, Nerem RM (1995b): Co-culture of endothelial cells and smooth muscle cells in a flow environment: An improved culture model of the vascular wall? *Cells Mat* 5:115–124

Zilla PP, Fasol RD, Deutsch M, eds. (1986): *Endothelialization of Vascular Grafts.* Basel: S Karger

10

NEW FRONTIERS IN TISSUE ENGINEERING: TISSUE ENGINEERED HEART VALVES

TOSHIHARU SHIN'OKA AND JOHN E. MAYER JR.

INTRODUCTION

Valvular heart disease is a significant cause of morbidity and mortality world wide (Rabago, 1987). In the United States, approximately 20,000 Americans die as a direct result of their valvular dysfunction every year. In 1991, 58,000 valve replacement operations were performed in this country (Brauswald, 1994). In pediatric patients all four cardiac valves may have abnormalities in association with various complex congenital lesions (Mullins, 1993).

Currently, valve replacement represents the most common mode of surgical therapy for the treatment of end-stage valvular heart disease. Valve replacement surgery is efficacious, and it has substantially reduced the morbidity and mortality associated with valvular dysfunction. However, in the pediatric population (Carpentier, 1983), valve reparative surgery is preferable because there are no ideal prosthetic valve for these smaller-size patients, and valve repair operations have gained an important role in the management of acquired heart disease in adults because of the shortcomings of current valve replacement devices.

The state-of-the-art valves used clinically include glutaraldehyde fixed xenograft valves, cryopreserved homograft valves, and mechanical valves (Table 1). While the currently used prosthetic valves function well, each has its inherent limitations. The major drawback of using a mechanical valve relates to the fact that it represents a foreign body, which is associated with risks of infection and thromboembolic complications. Therefore, use of a mechanical valve requires the administration of life-long anticoagulation with all of its associated morbidity (Cannegieter et al, 1994). Glutaralde-

Synthetic Biodegradable Polymer Scaffolds
Anthony Atala and David Mooney, Editors
Robert Langer & Joseph P. Vacanti, Associate Editors
© 1997 Birkhäuser Boston

Table 1. Limitations of Current State-of-the Art Valves

Mechanical Valve	Foreign Body Response
	Lack of Growth
	Mechanical Failure
	Need for Life-long Anticoagulation
	Thrombosis
Tissue Valve (Xenograft)	Foreign Body Response
	Lack of Growth
	Short durability
	Calcification
Allogenic Valve (Homograft)	Foreign Body Response
	Lack of Growth(?)
	Donor Organ Scarcity
	Rejection(?)

hyde-fixed xenograft valves are more biocompatible than their mechanical counterparts and, therefore, are not associated with the same risk of thromboembolic complications. The major disadvantage of the glutaraldehyde-fixed xenograft valve relates to its relative lack of durability. Most such valves require replacement within 10–15 years, and the durability is much less in younger patients. Cryopreserved homograft valves contain viable cells, but they are allografts and, therefore, are subject to rejection. The durability of homografts is also less than prosthetic valves. Nevertheless, homografts have the advantages of viability, better resisting infection and potential growth, although the latter remains debatable (Kirklin et al, 1993). In addition, the supply of cryopreserved homograft valves is severely limited by donor organ scarcity (Braunwald, 1992, Watts et al, 1976).

POTENTIAL BENEFITS OF AN AUTOLOGOUS TISSUE ENGINEERING APPROACH TO VALVULAR HEART DISEASE

Because of the limitations associated with the various valve substitutes, a series of experiments have been undertaken to determine if tissue engineering principles could be used to develop tissues suitable for replacement of all or part of a cardiac valve. We reasoned that the creation of an autologous tissue engineered heart valve would offer several theoretical advantages over the presently used valve substitutes. It was anticipated that an autologous bioprosthesis would be a living structure, and, therefore, it should be able to demonstrate the normal biological mechanisms for growth, repair, and development which theoretically would translate to a greater durability. A tissue engineered autologous leaflet would be completely biocompatible, with

Table 2. Theoretical Advantages of Autologous
Tissue Engineered Structure

No rejection
No donor organ scarcity
Viable structure and greater durability
No foreign body response
No need for long-term anticoagulation
Growth potential

---

minimal risk of infection and thromboembolic complications. It remained un-proven that such a leaflet could be engineered or that it would have any of these theoretically attractive capabilities (Table 2).

TISSUE ENGINEERED HEART VALVES: CURRENT STATUS

This project to develop tissue engineered valve replacement tissue is based on the observation that a normal heart valve is composed mainly of endo-thelial cells and myofibroblasts. Therefore, tissue for cell harvesting and in vitro expanding can be obtained from vascular structures that contain simi-lar cell components. In the following sections, the initial experiments which have demonstrated the feasibility of constructing a tissue engineered valve leaflet tissue are described.

Cell expansion technique

Twenty day old Dover lambs were sedated with ketamine, intubated, and placed on a mechanical ventilator. After induction of general anesthesia with diprivan a longitudinal incision was made in the left groin under sterile conditions. A 2 cm section of the femoral artery was resected and placed into sterile saline solution. The proximal and distal ends of the artery were ligated. After wound closure anesthesia was discontinued.

After harvesting, the tissue specimens were washed in Dulbecco's phos-phate buffered saline. In a laminar flow hood the tissues were minced into 1–2 mm^2 sections using sterile technique. The minced tissues were evenly distributed over 15 mm $\times$ 60 mm tissue culture plates and were allowed to air dry for twenty minutes. After the pieces were firmly attached to the tissue culture plates, medium was gently added taking care not to dislodge the explants. The medium consisted of Dulbecco's Modified Eagles media sup-plemented with 10% fetal calf and 1% antibiotic solution (L-glutamine 29.2 mg/ml, Penicillin G 1000 units/ml, and Streptomycin sulfate 10,000 µg/ml).

After 10–14 days, cells began to migrate off the explanted tissue and onto the tissue culture plates. Cell growth from the explants was observed after several days to form mixed cell populations of endothelial cells and fibroblasts. After 2–3 weeks, the mixed cell population grew into a confluent monolayer.

After the explanted cell cultures became confluent, they were serially passaged by trypsinization. Two ml of 0.25% trypsin/1 mM EDTA was added to each confluent tissue culture plate and incubated at 37°C for ten minutes. The plate was mechanically agitated facilitating the formation of a single cell suspension. Thirty ml of serum-containing media was added to inactivate the trypsin. The cell suspension was gently vortexed to create an even cell suspension which was divided equally among four 75 cm^2 vented tissue culture flasks. An additional 45 ml of media were added to each flask, which were incubated at 37°C, in a 10% CO_2 atmosphere. The cells grew into confluent monolayers after 7–10 days.

After reaching confluence, endothelial cells were fluorescently labeled in preparation for sorting. A 2 µg/ml labeling solution was prepared by adding 100 µg of fluorescently labeled acetylated low density lipoprotein (Dil-Ac-LDL) to 50 ml of media. The cultured cells were then labeled with an acetylated LDL marker which is selectively taken up by endothelial cells via the scavenger pathway. After a 24 hour incubation period, the cells were trypsinized and resuspended to create a single cell suspension consisting of labeled endothelial and unlabeled mixed fibroblast cells population. The labeled mixed-cell population was sorted using a fluorescent-activated cell-sorter (Bectin Dickenson, Mountainview, CA). Cells were sorted into LDL$^+$ and LDL$^-$ fractions based on their size and their fluorescence. The two cell populations were then separately cultured to obtain sufficient cell numbers for polymer seeding. The media was changed every seven days, and cell growth was assessed periodically. The cell populations were expanded by repeated passages. Cell viability was typically greater than 95%, as determined by trypan blue exclusion (Breuer, 1996).

Leaflet construction

In a typical seeding, 20×10^6 LDL$^-$ cells (myofibroblasts) were seeded onto a polymer mesh. This tissue scaffold was composed of a polyglactin woven mesh sandwiched between two nonwoven PGA mesh sheets which measured 3 cm by 3 cm in size and 3.2 mm in thickness. The matrix was greater than 95% porous before seeding. The scaffold was designed so that it would be biodegraded over a 6 to 8 week period (Ma, 1995). The cell/polymer constructs were placed into culture plates for one hour after cell seeding. After cell attachment, 50 cc of media were added. Culture

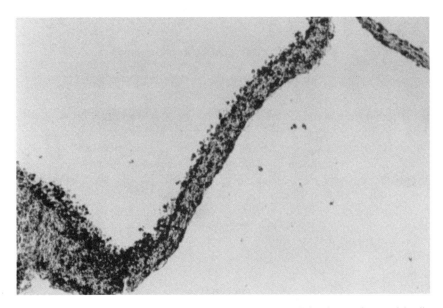

Figure 1. Immunohistochemical staining of Factor VIII in the engineered leaflet before implantation.

media were changed on a daily basis. After 7 days, the tissue engineered construct was seeded with additional two million of endothelial cells per cm², which formed a monolayer around the fibroblast-polymer construct (Figure 1).

Labeling for cell tracing

In order to determine whether cells seeded onto the polymer in vitro persisted after implantation, FM-DiI(F-6999) (Molecular Probes Inc, Eugene, OR) was used for irreversible cell labeling (Honig and Hume, 1986). Five hundred micrograms of FM-DiI were added to the cell suspension containing four million cells of a mixed cell population. The working solution was incubated in a 37°C water bath for 3 minutes, then transferred into ice for 1 minute before it was refrigerated at 4°C for another 14 minutes. The cell suspension was centrifuged and resuspended in Phosphate Buffered Solution. After another centrifugation, the cell pellet was resuspended in a warm culture medium. The cell suspension was centrifuged, collected, and then seeded onto the polymer sheet and cultured for one additional week prior to implantation.

Leaflet implantation

Pulmonary Valve Leaflet Replacement

Autologous tissue engineered leaflets were implanted into the same lambs from which the arterial wall cells had been previously harvested. In control animals, only polymer scaffold leaflets without cells were implanted. Anesthesia was induced with 30 mg/kg of ketamine and maintained with continuous infusion of 0.2 mg/kg/min of propofol. The chest was exposed through a left thoracotomy at the third intercostal space. Normothermic femoral arterial and right atrial cardiopulmonary bypass was established. With the heart beating, a longitudinal pulmonary arteriotomy was made. The right posterior pulmonary leaflet was excised and replaced with either a tissue-engineered valve leaflet or a plain polymer sheet using absorbable 5-0 PDS sutures (Figure 2). Tissue engineered valve leaflet has a unique advantage compared to the construction of other tissues or organs since the implanted cells are not limited by the development of vascular supply. Therefore, further in vivo cell growth can be expected (Shin'oka et al, 1996).

After 6 hours, 1, 6, 7, 9, and 11 weeks, the animals were sacrificed, and the implanted valve leaflets were examined functionally, histologically, biochemically, and biomechanically. 4-hydroxyproline assays were done for collagen content (Bergman and Loxley, 1963). Valve function was evalu-

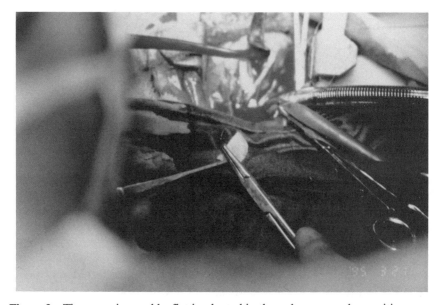

Figure 2. Tissue engineered leaflet implanted in the pulmonary valve position.

ated in vivo using Doppler echocardiography. Mechanical strength was evaluated in vitro with a Vitrodyne V-1000 mechanical tester. Factor VIII and elastin stains were done to verify histologically that endothelial cells and elastin were present, respectively (Booyse et al, 1974; Tawfik et al, 1994). Animals receiving leaflet made from polymer without cell seeding were sacrificed and examined in a similar fashion after 8 weeks. In the control animals, the acellular polymer leaflets were completely degraded leaving no residual leaflet tissue at eight weeks. The tissue engineered valve leaflet persisted in each animal in the experimental group (Figure 3). Echocardiographic examination of the pulmonary valve demonstrated no evidence of stenosis and trivial pulmonary regurgitation in the lambs receiving autologous leaflets (Figure 4). 4-hydroxyproline analysis of the constructs showed progressive increase in collagen content (Figure 5). There was a trend toward an increase in tensile strength over time (Figure 6). Immunohistochemical staining demonstrated elastin fibers in the matrix (Figure 7a) and Factor VIII on the surface of the leaflet (Figure 7b). The cell labeling experiments demonstrated that the cells on the leaflets had persisted from the in vitro seeding of the leaflets (Shin'oka et al, 1995).

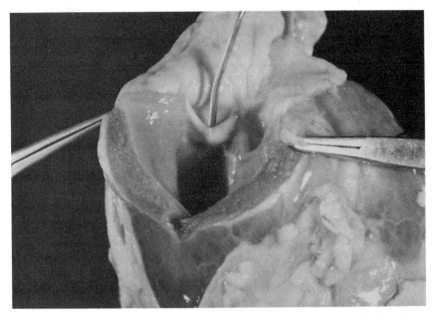

Figure 3. Gross morphology of tissue engineered leaflet 11 weeks after implantation.

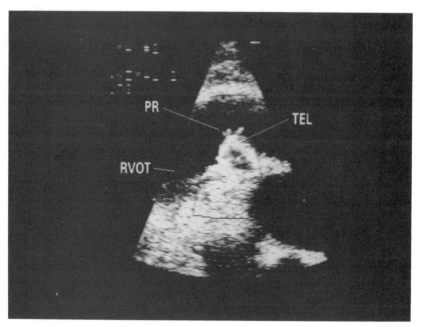

Figure 4. Echocardiography demonstrated minimal pulmonary regurgitation and no evidence of stenosis.

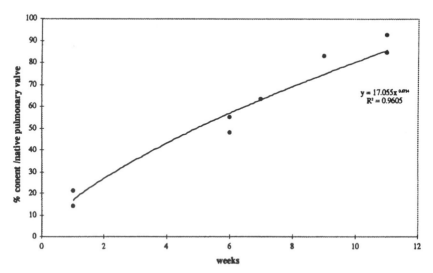

Figure 5. Collagen assay is expressed as % of native pulmonary valve collagen content. Note the gradual increase in collagen content of the engineered leaflet over the 11 week period.

$$\% \text{ Collagen} = \frac{\text{Tissue–engineered collagen content} \times 100}{\text{Native pulmonary leaflet collagen}}$$

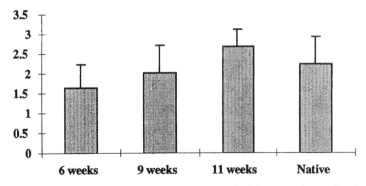

Figure 6. Mechanical testing demonstrated a gradual increase in maximal tensile strength over time. Note the similar mechanical properties between the engineered leaflets and native pulmonary leaflet 6 weeks after implantation.

Problems with current technique

There are still several limitations that must be overcome before the clinical application of these tissue engineered valve structures will be possible. A major limitation is the stiffness of the tissue engineered valve leaflets. The

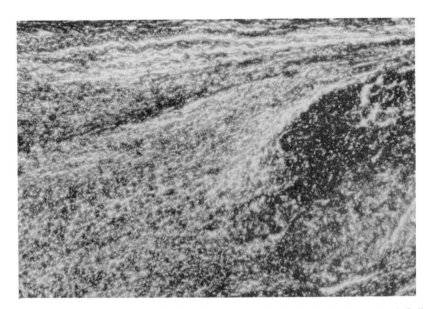

Figure 7a. Immunohistochemical staining shows elastin (red staining material) fibers between collagen fibers in cross-section of a valve construct

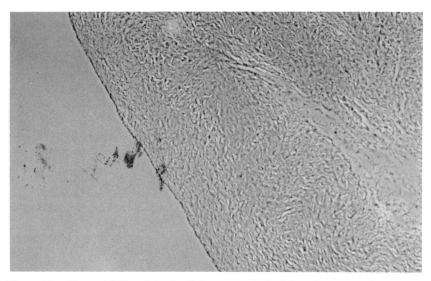

Figure 7b. Factor VIII staining(red) for the endothelial cells was positive in cells along the outer layer of the tissue engineered valve leaflet.

optimal duration of in vitro seeding remains to be determined because prolonged incubation results in the degradation of the bioabsorbable polymer which is then rendered unsuitable for suturing. On the other hand, the composite polymer remains too stiff and nonpliable if implanted prematurely.

In addition, the relatively nonpliable nature of the Vicryl mesh creates a logistic problem for the design of a whole trileaflet valve. At the time of this writing, we have attempted several unsuccessful tissue engineered whole valve implantations. The nonpliable valves inevitably result in the development of severe pulmonary stenosis if all three leaflets are replaced.

One additional issue relates to the maintenance of the composite polymer integrity under conditions of greater hemodynamic stress. All experiments to date have been carried out in the low pressure pulmonary circulation. We are currently exploring techniques by which the tissue engineered valvular tissue can be "conditioned" in vitro prior to implantation.

FUTURE DIRECTION

The engineering of autologous cardiovascular structure is still in an early stage of development, and there are numerous issues that remain to be investigated at the present time. However, we are encouraged that the construction of autologous valve tissue replacements seem to be feasible.

Active experimental research in this field is currently underway in this institution in a number of areas. We are currently investigating the longer-term durability of tissue engineered leaflets, their growth potential in growing animals, alternative sources of cells (vein cells, arterial cells, dermal fibroblasts, capillary cells, and allogenic cells), and variations in cell seeding technique (mixed cells seeding versus staged selective seeding). We are also attempting to make other tissue engineered cardiovascular structures such as pulmonary artery, femoral artery, and composite arterial conduits containing valve structures. In addition, work has started on the development of a flexible elastic polymer by our collaborators at the Massachusetts Institute of Technology. If they can be synthesized, these flexible polymers will likely have great benefits for tissue engineering, especially in the cardiovascular field. We believe that, based on these initial studies, a tissue engineering approach to the development of replacement cardiovascular structures is worthy of significant additional investigation.

References

Bergman I, Loxley R (1963): Two improved and simplified methods for the spectro-photometric determination of hydroxyproline. *Anal Chem* 35:1961–1965

Booyse F M, Sedlak B J, Rafelson M Jr (1975): Culture of arterial endothelial cells: Characterization and growth of bovine aortic cells. *Thromb Diath Haemorrh* 34:825–839

Braunwald E (1992): Valvular Heart Disease. In: *Heart Disease 4th edition,* Braunwald E ed. Philadelphia: Saunders Company

Breuer CK, Shinoka T, Tanel RE, Zund G, Mooney DJ, Ma PX, Miura T, Colan S, Langer R, Mayer JE, Vacanti JP (1996): Tissue engineering lamb heart valve leaflets. *Biotech Bioeng:* in press

Cannegieter SC, Rosendaal FR, Briet E (1994): Thromboembolic and bleeding complications in patients with mechanical heart valve prostheses. *Circulation* 89:635–641

Carpentier A (1983): Congenital malformations of the mitral valve. In: *Surgery for Congenital Heart Defects,* Stark J, de Levcal M, eds. London: Grune & Stratton

Heart and Stroke Facts: *1994 Statistical Supplement Framingham:* American Heart Association

Honig MG, Hume RI (1986): Fluorescent carbocyanine dyes allow neurons of identified origin to be studies in long term culture. *J Cell Biol* 103:171–187

Kirklin JK, Smith D, Novick W, Naftel DC, Kirklin JW, Pacifico A, Nanda NC, Helmcke FR, Bourge RC (1993): Long-term function of cryopreserved aortic homografts. *J Thorac Cardiovasc Surg* 106:154–166

Ma PX, Langer R (1995): Degradation structure and properties of fibrous nonwoven poly(glycolic acid) scaffolds for tissue engineering. In: *Polymers in Medicine and Pharmacy,* Mikos AG, Leong KW, Radomsky ML, Tamada JA, Yaszemski, eds. Pennsylvania: Materials Research society

Mullins CE (1993): Congenital valvular heart disease in pediatrics. In: *Valvular Heart Disease,* Zaibag MA, Duran CMG, eds. New York: Marcel Dekker

Rabago G (1987): A worldwide overview of valve usage. In: *Heart Valve Replacement & Future Trends in Cardiac Surgery,* Rabago G, ed. New York: Futura Publishing Company

Shinoka T, Breuer CK, Tanel RE, Zund G, Miura T, Ma PX, Langer R, Vacanti JP, Mayer JE (1995): Tissue engineering heart valves: Valve leaflet replacement study in a lamb model. *Ann Thorac Surg* 60:S513–516

Shinoka T, Ma PX, Shum-Tim D, Breuer CK, Cusick RA, Zund G, Langer R, Vacanti JP, Mayer JE (1996): Tissue Engineering Heart Valves: Autologous Valve Leaflet Replacement Study in a Lamb Model. *Circulation:* 94:SV164–168

Tawfik M, Malak, MD, Stephen C, Bell D (1994): Distribution of Fibrillin-containing microfibrils and elastin in human fetal membranes; A novel molecular basis for membrane elasticity. *Am J Obst Gynecol* 171:195–295

Watts LK, Duffy P, Field RB, Stafford EG, O'Brien MF (1976): Establishment of a viable homograft cardiac valve bank: A rapid method of determining homograft viability. *Ann Thorac Surg* 21:230–236

11

TISSUE ENGINEERING CARTILAGE AND BONE

YILIN CAO M.D., PhD, CLEMENTE IBARRA M.D., AND CHARLES A. VACANTI M.D.

INTRODUCTION

Tissue engineering is an emerging multidisciplinary field in which the material properties of synthetic compounds are manipulated to enable delivery of an aggregate of dissociated cells into a host in a manner that will result in the formation of new functional tissue. A high density of viable, dissociated functional cells can be seeded onto synthetic, biocompatible, biodegradable polymers of an appropriate chemical composition and physical configuration, which act as scaffolds. They allow diffusion of nutrients to the cells, as well as cell-to cell contact and can then be transplanted into animals. The purpose of this new field of research is to repair, replace, maintain, or enhance the function of a particular tissue or organ. Cells can be isolated from an individual, expanded in vitro and/or modified by gene therapy to replace a defective gene, and reimplanted.

A successful area of tissue engineering has been that of cartilage and bone. Autologous chondrocytes in culture have already started to be used in the clinical setting as means of "cell-therapy", to repair articular cartilage defects (Brittberg et al, 1994). Experimentally, tissue engineered cartilage has shown potential for application in different fields of surgery such as orthopaedics, plastic and reconstructive surgery, urology, and other medical specialties (Puelacher et al, 1994b,c; Vacanti et al 1992; 1994b,c). Congenital, traumatic, infectious, and neoplastic bone diseases that result in large bone defects are other areas in which tissue engineering of bone could offer an attractive alternative. The culture of osteoblasts and mesenchymal bone-re-

Synthetic Biodegradable Polymer Scaffolds
Anthony Atala and David Mooney, Editors
Robert Langer & Joseph P. Vacanti, Associate Editors
© 1997 Birkhäuser Boston

lated cells has also been found to be useful in the study of the biology of bone formation, and the effects of naturally or synthetically obtained growth factors such as transforming growth factor beta (TGFβ), and other members of its superfamily, such as bone morphogenetic proteins (BMPs) and other peptides (Kataoka and Urist, 1993).

During the last decade, researchers in our laboratory have been investigating the concept of using man-made, biodegradable scaffolds as delivery vehicles for cell transplantation. These polymer scaffolds allow cells to be delivered and immobilized in a given location and serve as a template for tissue development both before and after cell transplantation. The biodegradable polymer templates eventually resorb, avoiding a chronic foreign body response. Synthetic polymers can be synthesized reproducibly, are readily processed into devices of varying sizes and shapes, and have chemical and mechanical properties that can be tightly controlled and manipulated. The use of chemically engineered synthetic polymers allows very precise manipulation of the cell-delivery devices in regard to physical characteristics and mechanical properties of the materials, porosity, degradation times, and suitability for cell attachment. Growth factors and other peptides can also be attached to the side chains of the synthetic polymers to modify their properties or those of the cells seeded on them for the creation of new tissue.

Naturally occurring biomaterials (i.e., collagen) can also be used to produce these devices; however, natural biomaterials must be isolated from human or animal tissue and thus suffer large batch-to-batch variation. Natural materials also offer limited versatility in designing a device with specific properties such as mechanical strength and degradation time.

In this chapter we will attempt to review the current advances on the field of tissue engineering and discuss the engineering of bone and cartilage tissue we have been investigating. The approach of using biodegradable polymer templates to guide the development of transplanted cells into new tissue seems to be widely applicable. We hope that this will lead to the development of a core technology applicable to many tissue types needed not only for transplantation but also for surgical reconstruction.

TISSUE ENGINEERING OF CARTILAGE

Spontaneous repair of cartilage is very limited after injury or disease (Fuller and Ghadially, 1972; Ghadially et al, 1977; Mankin, 1982; Meacham, 1963). Development of early osteoarthritis (OA) is frequently the result of injured articular cartilage (Davis et al, 1989). Absent, defective, or injured cartilaginous structures throughout the body are difficult to reconstruct or repair, necessitating the use of prosthetic devices for these purposes. The use of synthetic materials in the reconstruction of cartilage structures presents several potential risks and complications. Infection, protrusion through the skin, breakage or loosening of the implants, and uncertain long-term immunologic interactions with the host can limit their indications and uses (Langer and Vacanti, 1993; Vacanti and Mikas, 1995; Vacanti and Upton, 1994). Cartilage obtained from remote sites in the same patient, and carved to the desired shape is commonly used to reconstruct cartilage structures such as the ear (Cocke, 1995; Destro and Speranzini, 1994; Mooney, 1995). However, secondary morbid sites are created, and neither aesthetically or functionally acceptable results are obtained with this technique. The use of allografts presents the problems of donor matching and tissue storage, scarcity of supply, increased susceptibility to infection, and possible increased risk of transmission of infectious agents (Asselmeier et al, 1993).

Different attempts have been made to repair or replace damaged cartilage with biologic substitutes. Perichondrium and periosteum have been used with limited success (Bruns et al, 1992; Tsai et al, 1992; Upton et al, 1981), although some reports suggest that continuous passive motion can stimulate the formation of cartilage when articular chondral lesions are repaired with periosteum (O'Driscoll and Salter, 1984; O'Driscoll et al, 1986). Exposing periosteum to growth factors such as TGF-beta 1, has also been reported to stimulate chondrogenesis by periosteal cells (Miura et al, 1994). Long term studies will be necessary to demonstrate that the cartilage formed by periosteal cells stimulated by growth factors will not degenerate and break down, as did the fibro-cartilage formed when perforations were drilled in subchondral bone in an attempt to repair a damaged articular surface. Autografts (Matsusue et al, 1993; Springfield, 1987) and allografts (Friedlander and Mankin, 1984; Mankin et al, 1987), which rely on creeping substitution by host cells for integration, are currently used to reconstruct osseous and osteochondral defects. Nevertheless, articular cartilage does not seem to incorporate to the surrounding tissue when transplanted as does bone. Allogeneic cartilage has shown temporary success when transplanted as part of a massive osteochondral graft, but degeneration overcomes the transplanted tissue over time, resulting in failure (Ostrum et al, 1994). Even though chondrocytes seem to survive transplantation better than other cell types, cartilage allografts usually undergo degenerative changes due to mechanical and biochemical factors (Bentley and Greer, 1971; Chesterman and Smith, 1968; Green, 1977). The use of autologous cells to repair an articular cartilage lesion or to replace a damaged or absent cartilaginous structure could address these problems in a more reasonable and lasting manner. In 1994 Brittberg et al described a technique to repair deep cartilage defects in human knee joints using cultured autologous chondrocytes in suspension. A chondral defect in a patient's knee was covered with a periosteal flap harvested from the anterior tibia during the same procedure, and chondrocytes suspended in culture medium were injected into the defect. Similar techniques to resurface injured joints using chondrocytes in suspension had previously been described in experimental studies by different authors, with varying degrees of success (Bentley and Greer, 1971; Chesterton and Smith, 1968; Green, 1977; Moskalewski, 1991; Ostrum et al, 1994). An update on the results presented by the same group during the annual meeting of the American Academy of Orthopaedic Surgeons in February of 1996 revealed somewhat discouraging, incomplete data that suggested similar results to experimental series previously described. Clinical trials are also being conducted currently using periosteal flaps exposed to TGF-beta 1 to resurface damaged articular cartilage. Long term follow-up of treated patients and a very precise characterization of the tissue formed in the defect will be necessary to demonstrate the nature and source of the cells within the repair tissue and the biochemical and biomechanical properties of their surrounding matrix.

The results obtained with the use of xenografts have not been promising either (Lipman et al, 1993). Nevertheless, matrix components of animal-derived tissue have been investigated and used experimentally as potential cell carriers.

In the 1970s, Green (1977) attempted to create cartilage by seeding isolated chondrocytes on demineralized bone, unsuccessfully. Cell-delivery devices had to be developed when the natural difficulties of using cells in suspension became evident. The systems had to allow cell delivery and survival at the desired site to produce cartilage. Investigators studied different naturally-derived carriers with varying rates of success. Wakitani et al (1989) and Kimura et al (1983) described the use of chondrocytes embedded in collagen gels to create cartilage. These gels not only permitted cell-delivery but allowed chondrocytes to maintain their phenotype during culture and avoid dedifferentiation observed during monolayer culture (Takigawa et al, 1987). Itay reported the use of fibrin glue as a vehicle in chondrocyte transplantation with limited ability to repair cartilage lesions (Itay et al, 1987). Cell behavior and survival in fibrin glue can limit its use (Homminga et al, 1993). Fibrin glue and collagen sponges have also been used to generate hyalinelike cartilage from perichondrium, with poor integration of the newly formed tissue to the surrounding cartilage (Upton et al, 1981). Agarose gels have been used as scaffolds to support chondrocytes for culture in vitro (Siltinger et al, 1994), and peptide stimulation of cells has also been studied as another means of creating new cartilage (Hsieh et al, 1996; Tesch et al, 1992; Wozney, 1988). Moreover, genetic engineering has recently been proposed as an alternative approach to resurfacing articular cartilage defects. Exogenous genes have been introduced in rabbit articular chondrocytes via retrovirus. The transfected cells were then allotransplanted to repair articular cartilage defects. This is believed to have potential in future retrovirally-mediated introduction of transgenes that can stimulate the synthesis of matrix components by the transfected cells and possibly by the surrounding native cells (Kang et al, 1996).

The scarcity of organs and tissues for transplantation stimulated the search for new alternatives. One of the alternatives proposed in the early 1980s, was the de novo creation of parenchymal tissues by tissue engineering. Employing similar methodology, Vacanti (1988) and Vacanti et al (1988) created new cartilage by combining cell isolation and tissue culture procedures with the use of synthetic, biocompatible, biodegradable materials. In 1991 Vacanti et al were able to create new hyaline cartilage subcutaneously in athymic mice, by transplanting bovine chondrocytes seeded onto unbraided biodegradable suture materials. The cells attached to the fibers while creating their own supporting matrix. The temporary scaffold provided by the suture fibers allowed for diffusion of nutrients and waste products and began to resorb at the same time that a natural supporting matrix was being formed by the cells. In this manner, the cells were able to bridge the interfiber spaces with extracellular matrix while attaching to the polymer and to each other, forming new tissue. The generation of cartilage was evaluated at different times by gross and histologic examination, using hematoxylin and eosin (H&E) (Figure 1), safranin-O, and alcian blue stains. Cells were also labeled in vitro with BrdU or fluorescein (chloromethyl derivative of amino and hydroxycoumarin) before implantation to determine whether the tissue generated had been formed by the transplanted cells. After the first successful attempts using unbraided sutures of polyglactin 910 (Vicryl, Ethicon, Somerville, NJ) and polyglycolic acid (PGA)(Dexon, Davis & Geck, Danbury, CT), different polymer configurations were utilized. PGA fibers configured as a nonwoven mesh of approximately 100µm thick, with a fiber diameter of 14–15µm and interfiber spaces of 150–200 µm, and copolymeric structures combining PGA and poly-L-lactic acid (PLLA) were employed to create cartilage constructs in predetermined shapes (Kim et al, 1994). Shape, size, time of degradation, and affinity for cell attachment of the polymer scaffolds could be controlled very precisely by manipulating the types and amounts of polymers used. The optimal cell concentrations to create cartilage from cell-polymer constructs were determined (Puelacher et al, 1994a). Cell survival studies were performed demonstrating the capacity of chondrocytes to maintain functional activity after being stored in a refrigerator at 4°C in an appropriate culture media for up to 30 days (Kim et al, 1993). The ability of chondrocytes to multiply and form new cartilage was also demonstrated with the use of cells isolated from articular cartilage obtained from a 100-year-old patient (Vacanti et al, 1994a).

Injectable polymers were also studied as cell delivery devices to create new cartilage (Paige and Vacanti, 1995; Paige et al, 1995). Alginate hydrogel solutions were used to suspend chondrocytes. The cell-polymer suspension

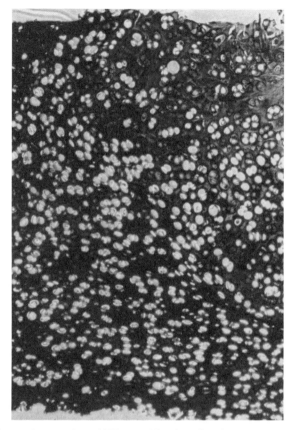

Figure 1. Photomicrograph at 20X magnification showing a segment of a 6 week specimen of tissue engineered cartilage stained with toluidine blue. Notice the morphology of abundant cells within a poteoglycan-rich matrix as suggested by the strongly positive stain. The cells have not completely acquired a columnar arrangement pattern, frequently observed in more mature specimens and normal articular cartilage.

was then polymerized with calcium and injected through a syringe into plastic molds to form discs which were implanted into subcutaneous pockets in athymic mice. The chondrocyte-calcium alginate gel was also injected directly into the subcutaneous space of the dorsum of nude mice. Cartilage was formed after 8 weeks of implantation. Finally, very specific applications of the new tissue-engineered cartilage have been tested. Nasoseptal implants and a temporomandibular joint disc have been designed using tissue engineered cartilage (Puelacher et al, 1994b,c). Cartilage in the shape of a human ear has been successfully obtained (Vacanti et al, 1992), and a similar model of vascularized autologous cartilage covered with skin is currently being developed for use as a pedicled graft in ear reconstruction in a rabbit model (Cao and Vacanti, 1996). Cartilaginous tubes lined with respiratory epithelium have been engineered for tracheal replacement (Ghadially et al, 1977), and the potential for joint resurfacing has been accomplished in a rabbit model with tissue engineered cartilage (Vacanti et al, 1994c). Recently, authors have reported to have successfully repaired articular cartilage defects in rabbits using similar cell-delivery devices and mesenchymal stem cells (Grande et al, 1995).

The potential applications of tissue-engineered cartilage are very numerous. The use of synthetic biodegradable polymers combined with cell stimulation and tissue formation enhancement by growth factors has already shown potential to improve the generation of cartilage (Zimber et al, 1995). Combining these resources to engineer cartilage in vitro (Freed et al, 1993a), with the aid of bioreactors (Duke et al, 1993; Freed et al, 1993b; Goodwin et al, 1993; Klement and Spooner, 1993; 1994; Sittinger et al, 1994), the possibility to generate tissue engineered cartilage in the laboratory from autologous cells, to treat injured or diseased articular cartilage or to reconstruct cartilaginous structures throughout the body, could very soon become a common and widely used procedure.

Tissue Engineering of Bone

Reconstruction of bone defects, regardless of their origin, remains a difficult and controversial problem for the orthopaedic and reconstructive surgeon. Many techniques for replacing the structure and function of bone lost by trauma or disease have been used. The ideal reconstruction in all cases would replace the bone defect with like (autogenous) tissue. However, the supply of autograft tissue is usually limited, and the harvesting procedures can lead to the creation of secondary morbid sites and potential postoperative complications. Graft tissue is also occasionally not suitable for the required reconstruction because of poor tissue quality or extreme difficulty shaping the graft. Allografts can be used in some cases, but problems associated with donor matching, integration, and tissue storage, coupled with potential complications related to blood-borne products, limit their indications and uses (Asselmeier et al, 1993; Mankin et al, 1987). Alloplastic materials are another alternative, but they can show increased susceptibility or risk of infection and/or extrusion and an uncertain long-term interaction with the host's physiology.

Organic and inorganic bone substitutes have also been used alone and in conjunction with either demineralized bone or autogenous bone grafts. Mesenchymal cell differentiation into bone has been achieved by means of polypeptide stimulation, demineralized bone powder, or both (Constantino et al, 1992; Dahlen et al, 1991; Gatti et al, 1990; Kataoka and Urist, 1993; Krukowski et al, 1990; Mulliken and Gowacki, 1980; Roux et al, 1988; Rozema et al, 1990; Thaller et al, 1993; Yukna, 1990). More recently, living cells have been delivered on alloplastic implants to produce bone (Ono et al, 1995). Cartilage growth, whether achieved by use of cell suspensions alone, perichondrium-derived cells attached to naturally occurring matrices, or peptide stimulation, had been described only in association with and confined to a defect in cartilage or its underlying bone. The search for new prosthetic materials continues today, primarily because at present none are exempt from significant potential problems. Whenever damaged bone is replaced by synthetic materials, a significant amount of adjacent bone has to be surgically removed, and there has always been a problem at the interface of the prosthetic material and the underlying bone.

Using similar techniques and applying the same principles to generate neocartilage, Vacanti et al (1993) engineered new bone by seeding bovine periosteal cells onto sheets of nonwoven PGA mesh and implanting the cell-polymer constructs subcutaneously in nude mice (Ono et al, 1995). Cells shed from periosteum explants obtained under sterile conditions from newborn calf forelimbs were collected in tissue culture dishes. Once the cells had formed a monolayer on the tissue culture dish, they were seeded onto the polymer by wiping the bottom of the dish with sterile sheets of PGA. The cells attached readily to the polymer as assessed by phase contrast microscopy. The cell-polymer constructs were cultured in vitro for an additional week until the polymer was coated with several layers of periosteal cells. Metabolic activity of the cells was determined before subcutaneous implantation of the cell-polymer constructs into athymic mice. Positive immunohistochemical staining for osteocalcin, a bone-specific protein, in the supernatant from the culture media confirmed the presence of functioning osteoblasts. Constructs were then implanted in subcutaneous pockets in the dorsum of nude mice and then harvested at different time points for gross and histologic examination. By 6 weeks, the constructs showed gross and microscopic appearance of cartilage with focal areas of vascular invasion and bone formation. Cartilage was observed only during the first few weeks of in vivo implantation, with subsequent transformation into mature, organized bone as determined by gross examination and histologic evaluation using H&E stains. After 10 weeks in vivo, the specimens showed bone morphology, with marked vascular proliferation. Areas of tissue undergoing an endochondral ossificationlike process and small islands of cartilage were still present. The newly formed bone contained cellular elements of bone marrow (Figure 2). Polymers seeded with chondrocytes isolated by enzymatic digestion from bovine articular cartilage generated only cartilage, which did not progress to become bone. This confirmed the observations that bone was generated only when periosteal cells were used.

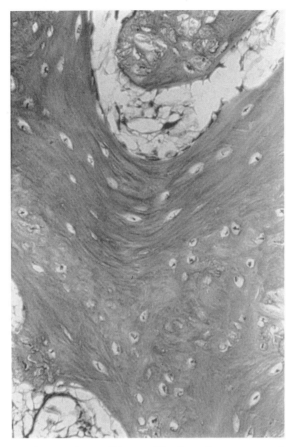

Figure 2. Photomicrograph at 40X magnification of a segment of a specimen of tissue-engineered bone (stained with H&E), obtained by subcutaneous injection of bovine periosteal cells suspended in calicium alginate, in the dorsum of a nude mouse for 12 weeks.

Applications

Reconstruction of Cranial Defects in Rats

Chondrocyte-PGA mesh constructs as well as osteoblast-polymer constructs have been successfully used to repair large structural defects created in the craniums of athymic rats. After the cranial periosteum had been excised, forty-eight bicortical defects, approximately two centimeters square, were made in the parietal, frontal, and temporal bones of athymic rats with a dremel drill. All defects extended to underlying dura and were irrigated with normal saline to remove any residual bone powder. Twenty-five percent of the defects were filled with chondrocyte-polymer constructs, 25% of the defects were filled with polymers seeded with osteoblasts isolated from periosteum. Another 25% of the defects were filled with polymer alone (not seeded with cells), while the final 25% of the defects were not reconstructed. Rats were sacrificed at 3 weeks, 6 weeks, 9 weeks, and 12 weeks. The site of defects was studied grossly and histologically using hematoxylin and eosin stains. No evidence of repair was observed in any defects filled with polymer alone or with nothing at all. At 3 weeks, defects filled with polymers containing periosteal cells and chondrocytes were virtually identical both grossly and histologically, with new cartilage spanning the defect. Six weeks after implantation, defects filled with periosteal cells showed evidence of what appeared to be "cartilage" undergoing some ossification, while those filled with chondrocytes demonstrated cartilage alone. In specimens implanted for 9 and 12 weeks, only cartilage was present in defects filled with polymers seeded with chondrocytes, while those filled with polymers seeded with periosteal cells had been repaired with new organized bone growth. These findings in conjunction with those from the subcutaneous implantation studies suggest that hyaline cartilage is a distinct tissue from bone and not the same as what appears to be cartilage on the developmental way to becoming bone.

More recently, we found similar bone and cartilage growth patterns in bone defects created in the femoral shafts of nude rats. We resected 2 cm long segments of the midfemoral shafts of nude mice after applying bridge plates. Defects were filled, in a manner similar to that described with the cranial defects, with either polymer seeded with chondrocytes or periosteal cells. Preliminary results have demonstrated "cartilage" bridging the defects four weeks after being filled with chondrocytes or periosteal cells on polymers. Examination of specimens implanted for longer periods of time revealed presence of cartilage in all defects repaired with polymers seeded with chondrocytes and new bone formation in only those defects filled with polymers seeded with periosteal cells.

Vascularized bone

Repair of skeletal defects with vascularized bone grafts has many advantages over nonvascularized grafts, but the availability of these grafts is extremely limited. A study was designed to determine whether new vascularized bone could be engineered by transplantation of osteoblasts around existing vascular pedicles using biodegradable polymers as cell delivery devices. Cells isolated from the periosteum of newborn bovine humerus were seeded onto polymer scaffolds as described above. After maintenance in vitro for 2 weeks, they were implanted around the right femoral vessels of athymic rats. Rats were sacrificed after six and nine weeks. New bone formation was evident in 10 of the 12 implants. At six weeks, the tissue was primarily composed of what appeared both grossly and histologically to be cartilage enveloping small islands of osteoid. The degree of osteoid and bone formation progressed with time as blood vessels invaded the tissue. This tissue ultimately underwent morphogenesis to become organized trabeculated bone engrafted on a vascular pedicle. We then demonstrated the ability of the vascularized bone to be transferred as a pedicle graft, which remained viable.

Midshaft bone defects created in the femur of nude rats and fixed with miniplates and screws to maintain the bone gap were filled with cell-polymer constructs in similar fashion. Radiographic controls showed evidence of bone formation and signs of bone healing in the periosteal cell-polymer group. Animals in which chondrocytes on polymer, plain polymer, or nothing had been used to fill the bone gap did not show any radiographic signs of bone formation. After 24 weeks, the miniplates were surgically removed. The bone defects filled with periosteal cells showed complete bone healing with exuberant callus formation covering the miniplates. Bone defects filled with chondrocytes on polymer showed cartilage formation filling up the gap. This explained the difference in radiographic appearance in the animals of these two groups. Finally, the bone gaps that had been filled with plain polymer or nothing showed atrophic nonunions. In some of the animals of these two groups the miniplates had broken.

Tissue Engineered Bone and Cartilage Composite Structures

Bone-cartilage composite structures have also been engineered by selectively seeding chondrocytes and periosteal (osteoblastic) cells on polymers or by suturing together cell-polymer scaffolds seeded with either chondrocytes or periosteal cells. In each construct only one of the two cell types was labeled with a fluorescent dye in vitro before implantation into animals for up to 16 weeks (Freed et al, 1993b). Early specimens showed formation of cartilage only. Over time, new bone and cartilage was demonstrated grossly and histologically. Bone developed exclusively on the side of the polymer construct originally seeded with periosteal cells while only cartilage remained on the opposite side of the construct with no evidence of bone formation (Figure 3). This was confirmed by detection of the specific labeled cells confined to one side of the specimen. A distinct bone-cartilage interface was formed and could clearly be identified by the presence of nylon suture material used to mark the area. This experiment further demonstrated that periosteal or osteoblastic cells cultured in vivo first generate tissue similar to cartilage, then eventually give rise to mature bone through

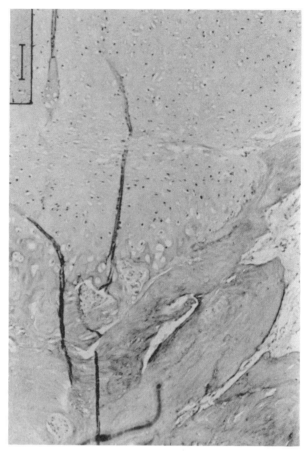

Figure 3. Photomicrograph at 20X magnification of a tissue engineered composite of bone (top) and cartilage (bottom) stained with H&E. Notice the very distinct interface between the 2 tissues. The black lines that cross the specimen are artifacts that appeared during preparation of the specimen.

an endochondral ossificationlike pathway, whereas chondrocytes implanted in similar fashion evolve into mature cartilage (Asselmeier et al, 1993).

Potential areas of future research might involve attempts to combine some of the strategies used by different researchers such as the use of new synthetic biodegradable polymer scaffolds with improved biomechanical properties (Yaszemski et al, 1995) and the use of autologous cells cultured and expanded in the presence of growth factors. It was necessary to elucidate the normal biology of isolated cells without exogenous stimuli prior to comparing the biologic effect of peptides when applied to tissue engineering. The stimulation and enhancement of cell growth and tissue formation with growth factors in vitro, with the aid of bioreactors (Wakitani et al,

1989), should allow for more controlled study and development of tissues and organs for transplantation.

In summary, tissue engineering has shown great potential in generating tissues such as cartilage and bone in the laboratory. With the advances shown by current results of different investigators in the field, and by perfecting the techniques, clinical trials for different human applications of tissue engneered bone and cartilage could soon be initiated.

References

Asselmeier MA, Caspari RB, Bottenfield S (1993): A review of allograft processing and sterilization techniques and their role in transmission of human immunodeficiency virus. *Am J Sports Med* 21: 170–75

Bentley G, Greer RG III (1971): Homotransplantation of isolated epiphyseal and articular cartilage chondrocytes into joint surfaces of rabbits. *Nature* 230: 385

Brittberg M, Lindahl A, Nilsson A, et al (1994): Treatment of deep cartilage defects in the knee with autologous chondrocyte transplantation. *NEJM* 331: 889–95

Bruns J, Kersten P, Lierse W, et al (1992): Autologous rib perichondrial grafts in experimentally induced osteochondral lesions in the sheep knee joint: Morphological results. *Virchows Archiv A Pathol Anat* 421: 1–8

Cao YL, Vacanti JP (1996): Unpublished data

Chesterman PJ, Smith AU (1968): Simon SR, ed. Homotransplantation of articular cartilage and isolated chondrocytes: An experimental study in rabbits. *J Bone Joint Surg* 50B: 184–97

Cocke WM Jr (1995): Radial composite chondrocutaneous flap for ear reconstruction. *Am Surgeon* 61: 347–9

Costantino PD, Friedman CD, Jones K, et al (1992): Experimental hydroxyapatite cement cranioplasty. *Plast Reconst Surg* 90: 174–91

Dahlin C, Alberius P, Linde A (1991): Osteopromotion for cranioplasty: An experimental study in rats using a membrane technique. *J Neurosurg* 74: 487–91

Davis MA, Ettinger WH, Neuhaus JM, et al (1989): The association of knee injury and obesity with unilateral and bilateral osteoarthritis of the knee. *Am J Epidemiol* 130: 278–88

Destro MW, Speranzini MB (1994): Total reconstruction of the auricle after traumatic amputation. *Plastic Reconstr Surg* 94: 859–64

Duke PJ, Daane EL, Montufar-Solis D (1993): Studies of chondrogenesis in rotating systems. *J Cell Biochem* 51: 274–82

Freed LE, Marquis JC, Nohria A, et al (1993a): Neocartilage formation in vitro and in vivo using cells cultured on synthetic biodegradable polymers. *J Biomed Mat Res* 27: 11–23

Freed LE, Vunjak-Novakovic G, Langer R (1993b): Cultivation of cell-polymer cartilage implants in bioreactors. *J Cell Biochem* 51: 257–64

Friedlander GE, Mankin HJ (1984): Transplantation of osteochondral allografts. *Ann Rev Med* 35: 311–24

Fuller JA, Ghadially FN (1972): Ultrastructural observations on surgically produced partial-thickness defects in articular cartilage. *Clin Orthop* 86: 193–205

Gatti AM, Zaffe D, Poli GP (1990): Behaviour of tircalcium phosphate and hydroxyapatite granules in sheep bone defects. *Biomaterials* 11: 513–17

Ghadially FN, Thomas I, Oryschak AF, et al (1977): Long-term results of superficial defects in articular cartilage: A scanning electron-microscope study. *J Pathol* 121: 213–17

Goodwin TJ, Prewett TL, Wolf DA, et al (1993): Reduced shear stress: A major component in the ability of mammalian tissue to form three-dimensional assemblies in simulated microgravity. *J Cell Biochem* 51: 301–11

Grande DA, Southerland SS, Manji R, et al (1995): Repair of Articular Cartilage Defects Using Mesenchymal Stem Cells. *Tissue Engin* 1: 345–354

Green WT Jr (1977): Articular cartilage repair: Behavior of rabbit chondrocytes durng tissue culture and subsequent allografting. *Clin Orthop* 124: 237

Homminga GN, Buma P, Koot HWJ, (1993): Chondrocyte behavior in fibrin glue. *Acta Orthop Scand* 64: 441–45

Hsieh PC, Thanapipatsiri S, Anderson P, et al (1996): Presented at the 42nd annual meeting of the Orthopaedic Research Society, Atlanta, GA, february, 1996

Itay S, Abramovici A, Nevo Z (1987): Use of cultured embryonal chick epiphyseal chondrocytes as grafts for defects in chick articular cartilage. *Clin Orthop* 220: 284–303

Kang R, Marui T, Nita IM, et al (1996): Gene therapy for full thickness articular cartilage defects. Presented at the 42nd annual meeting of the Orthopaedic Research Society, Atlanta, GA

Kataoka H, Urist MR (1993): Transplant of bone marrow and muscle-derived connective tissue cultures in diffusion chambers for bioassay of bone morphogenetic protein. *Clin Orthop* 286: 262–70

Kim WS, Vacanti JP, Upton J, et al (1993): *Potential of Cold-preserved Chondrocytes for Cartilage Reconstruction.* Plastic Surgery Research Council

Kim WS, Vacanti JP, Cima L, et al (1994): Cartilage engineered in predetermined shapes employing cell transplantation on synthetic biodegradable polymers. *Plast Reconstr Surg* 94: 233–37

Kimura T, Yasui N, Oshawa S, et al (1983): Chondrocytes embedded in collagen gels maintain cartilage phenotype during long-term cultures. *Clin Orthop* 186: 231–39

Klement BJ, Spooner BS (1993): Utilization of microgravity bioreactors for differentiation of mammalian skeletal tissue. *J Cell Biochem* 51: 252–56

Klement BJ, Spooner BS (1994): Premetatarsal skeletal development in tissue culture at unit- and microgravity. *J Exp Zool* 269: 230–41

Krukowski M, Shively RA, Osdoby P, et al (1990): Stimulation of craniofacial and intramedullary bone formation by negatively charged beads. *J Oral Maxillofac Surg* 48: 468–75

Langer R, Vacanti JP (1993): Tissue engineering. *Science* 260: 920–6

Lipman JM, McDevitt CA, Sokoloff A (1983): Xenografts of articular chondrocytes in the nude mouse. *Calcif Tissue Int* 35: 767

Mankin HJ (1982): Current concepts review. The response of articular cartilage to mechanical injury. *J Bone Joint Surg* 64A: 460–66

Mankin HJ, Gebhardt MC, Tomford WW (1987): The use of frozen cadaveric allografts in the management of patients with bone tumors of the extremities. *Orthop Clin N Am* 18: 275–89

Matsusue Y, Yamamoto T, Hama H (1993): Arthroscopic multiple osteochondral transplantation to the chondral defect in the knee associated with anterior cruciate ligament disruption. *Arthroscopy* 9: 318–21

Meachim G (1963): The effect of scarifcation on articular cartilage in the rabbit. *J Bone Joint Surg* 45B: 150–161

Miura Y, Fitzsimmons JS, Commisso CN, et al (1994): Enhancement of periosteal chondrogenesis in vitro. Dose response from transforming growth factor beta 1 (TGF-beta 1). *Clin Orthop Rel Res* 301: 271–80

Mooney KM (1995): External ear reconstruction with autogenous rib cartilage. *Plastic Reconstr Surg* 15: 92–7

Moskalewski S (1991): Transplantation of isolated chondrocytes. *Clin Orthop* 272: 16–20

Mulliken JB, Gowacki J (1980): Induced osteogenesis for repair and construction in the craniofacial region. *Plast Reconstr Surg* 65: 553–60

O'Driscoll SW, Salter RB (1984): The induction of neochondrogenesis in free intra-articular periosteal autografts under the influence of continuous passive motion. An experimental investigation in the rabbit. *J Bone Joint Surg* 66: 1248–57

O'Driscoll SW, Keeley FW, Salter RB (1986): The chondrogenic potential of free autogenous periosteal grafts for biological resurfacing of major full-thickness defects in joint surfaces under the influence of continuous passive motion. An experimental investigation in the rabbit. *J Bone Joint Surg* 68: 1017–35

Ono I, Gunji H, Suda K, et al (1995): Bone induction of hydroxyapatite combined with bone morphogenic protein and covered with periosteum. *Plas Reconstr Surg* 95: 1265–72

Ostrum RF, Chao EYS, Bassett CAL, et al (1994): Bone injury, regeneration and repair. In: Orthopaedic Basic Science American Academy of Orthopaedic Surgeons

Paige KY, Vacanti CA (1995): Engineering new tissue: Formation of neo-cartilage. *Tissue Engin* 1: 97–106

Paige KT, Cima LG, Yaremchuk MJ, et al (1995): Injectable cartilage. *Plas Reconstr Surg* 96: 1390–98

Puelacher WC, Kim SW, Vacanti JP (1994): Tissue engineered growth of cartilage: The effect of varying the concentration of chondrocytes seeded onto synthetic polymer matrices. *Oral Maxillofac Surg* 23: 49–53

Puelacher WC, Mooney D, Langer R, et al (1994b): Design of nasoseptal cartilage replacements synthesized from biodegradable polymers and chondrocytes. *Biomaterials* 15: 774–78

Puelacher WC, Wisser J, Vacanti CA, et al (1994c): Temporomandibular joint disc replacement made by tissue-engineered growth of cartilage. *J Oral Maxillofac Surg* 52: 1172–77

Roux FX, Brasnu D, Loty B, et al (1988): Madreporic coral: A new bone graft substitute for cranial surgery. *J Neurosurg* 69: 510–13

Rozema FR, Bos RR, Pennings AJ, et al (1990): Poly(L-lactide) implants in repair of defects of the orbital floor: An animal study. *J Oral Maxillofac Surg* 48: 305–9

Sittinger M, Bujia J, Minuth WW (1994): Engineering of cartilage tissue using bioresorbable polymer carriers in perfusion culture. *Biomaterials* 15: 451–56

Springfield DS (1987): Massive autogenous bone grafts. *Orthop Clin N Am* 18: 249–56

Takigawa M, Shirai E, Fukuo K, et al (1987): Chondrocytes dedifferentiated by serial monolayer culture from cartilage nodules in nude mice. *Bone Mineral* 2: 449–62

Tesch GH, Handley CJ, Cornell HJ, et al (1992): Effect of free and bound insuline-like growth factors on proteoglycan metabolism in articular cartilage explants. *J Orthop Res* 10: 14–22

Thaller SR, Hoyt J, Borjeson K et al (1993): Reconstruction of calvarial defects with organic bovine bone mineral (Bio-Oss) in a rabbit model. *J Craniofac Surg* 4: 9–84

Tsai CL, Liu TK, Fu SL, et al (1992): Preliminary study of cartilage repair with autologous periosteum and fibrin adhesive system. *J Formosa Med Assoc* 91: S239–45

Upton J, Sohn SA, Glowacki J (1981): New cartilage derived from transplanted perichondrium: What is it? *Plast Reconstr Surg* 68: 166–74

Vacanti CA, Mikos AG (1995): Letter from the editors. *Tissue Eng* 1: 1–2

Vacanti CA, Upton J (1994): Tissue engineered morphogenesis of cartilage and bone by means of cell transplantation using synthetic biodegradable polymer matrices. *Clinics Plastic Surg* 21: 445–62

Vacanti CA, Langer R, Schloo B, et al (1991): Synthetic polymers seeded with chondrocytes provide a template for new cartilage formation. *Plast Reconstr Surg* 87: 753–59

Vacanti CA, Cima LG, Ratkowski D, et al (1992): Tissue engineered growth of new cartilage in the shape of a human ear using synthetic polymers seeded with chondrocytes. *Mat Res Soc Symp Proc* 252: 367–73

Vacanti CA, Kim WS, Mooney D (1993): Tissue engineered composites of bone and cartilage using synthetic polymers seeded with two cell types. *Orthopaed Trans* 18: 276

Vacanti CA, Cao YL, Upton J, Vacanti JP (1994a): Neo-cartilage generated from chondrocytes isolated from 100-year-old human cartilage. *Transplan Proc* 2: 3434–35

Vacanti CA, Kim WS, Schloo B, et al (1994b): Joint resurfacing with cartilage grown in situ from cell-polymer structures. *Am J Sports Med* 22: 485–88

Vacanti CA, Paige KT, Kim WS (1994c): Experimental tracheal replacement using tissue engineered cartilage. *J Pediatr Surg* 29: 201–205

Vacanti JP (1988): Beyond transplantation. *Arch Surg* 123: 545–49

Vacanti JP, Morse MA, Saltzman WM (1988): Selective cell transplantation using bioabsorbable artificial polymers as matrices. *J Pediatr Surg* 23: 3

Wakitani S, Kimura T, Hirooka A, et al (1989): Repair of rabbit articular cartilage surfaces with allograft chondrocytes embedded in collagen gel. *J Bone Joint Surg* 63B: 529

Wozney JM (1988): Novel regulators of bone formation. Molecular clones and activities. *Science* 242: 1528

Yaszemski MJ, Payne RG, Hayes WC (1995): The ingrowth of new bone tissue and initial mechanical properties of a degrading polymeric composite scaffold. *Tissue Engin* 1: 41–52

Yukna RA (1990): Polymer grafts in human periodontal osseous defects. *J Periodont* 61: 633–42

Zimber MP, Tong B, Dunkelman N, et al (1995): TGF-beta promotes the growth of bovine chondrocytes in monolayer culture and the formation of cartilage tissue in three dimensional scaffolds. *Tissue Engin* 1: 289–300

12

TISSUE ENGINEERING OF BONE

REBEKAH D. BOSTROM AND ANTONIOS G. MIKOS

INTRODUCTION

Bone is a dynamic, highly vascularized tissue with the unique capacity to heal and to remodel depending on line of stress (Buckwalter et al, 1995ab). It exhibits the unlikely combination of high compressive strength and tensile strength due to the composite of calcium phosphate salts (hydroxyapatite) and collagen, respectively (Yaszemski et al, 1996a). It is difficult to find materials to mimic such a complex system when filling bone defects. However, current research capitalizes on the dynamic properties of bone by providing a biodegradable scaffold to guide healing.

Normal fracture healing of bone frequently requires temporary immobilization of the area but no other intervention. The normal healing process results in the reformation of organized, unscarred bone tissue. Certain situations may arise in which bone tissue does not heal completely; such a fracture is termed a nonunion and can be due to fracture comminution or bone loss, infection, loss of blood supply, disease, or inadequate fracture management (Yaszemski et al, 1995). In these situations, the patient suffers from the loss of mechanical support of the body's tissues and loss of muscle attachment sites. Such skeletal defects, thus, require therapeutic intervention to restore the bone to its proper form.

Therapies can be categorized into essentially two groups. Permanent replacement of the bone with a foreign material is a commonly used strategy. Bone cement for injectable filling of irregularly shaped skeletal defects and prefabricated metals for hip replacement are two bone replacement materials that lend high mechanical support. However, when a very strong material is used in bone tissue replacement, it absorbs the stresses of daily activity

Synthetic Biodegradable Polymer Scaffolds
Anthony Atala and David Mooney, Editors
Robert Langer & Joseph P. Vacanti, Associate Editors
© 1997 Birkhäuser Boston

instead of the bone; this phenomenon is called stress shielding. Since bone tissue normally maintains its strength in response to stress, the area surrounding the replacement material subsequently experiences bone loss (Bobyn et al, 1992). In addition, bone cement, which is formed by mixing prepolymerized poly(methyl methacrylate) (PMMA) powder with methyl methacrylate (MMA) monomer and initiator to form a moldable putty that hardens in a matter of minutes, has been associated with increased tendency for infections at the surgical site (Petty et al, 1988).

The second therapy involves the ultimate restoration of bone tissue at the defect site. Two well-established filler materials for this are autografts, bone tissue harvested from other areas of the patient, and allografts, bone harvested from cadavers. These provide a scaffold upon which the patient's adjacent bone tissue can invade, lay down new extracellular matrix (ECM), and remodel. In fact, the process of incorporation of transplanted bone is similar to normal bone regeneration after fracture or during growth (Yaszemski et al, 1996a). Any such material that provides a scaffold for bone tissue formation is termed osteoconductive.

Despite its widespread use, allograft transplantation is associated with several risks. Unprocessed allografts may carry diseases, such as hepatitis and AIDS or may be rejected by the immune system. While processed allograft is widely used as an osteoconductive material, it is believed that during processing allograft loses the factors that would induce bone tissue formation where and when it would not normally grow; that is, it may no longer be osteoinductive.

Problems associated with autografts are limited supply and donor site morbidity. Leaving aside these serious disadvantages, autograft satisfies the three requirements necessary for bone regeneration: it is osteoconductive, osteoinductive, and contains osteogenic cells (Gazdag et al, 1995; Yaszemski et al, 1996a). There is also no fear of rejection or transfer of disease.

Current research in bone tissue engineering uses cancellous (trabecular) autograft bone as the so-called gold standard to develop synthetic, porous, degradable scaffolds to fill defects (Gazdag et al, 1995). Since the scaffolds are degradable, bone tissue will invade the area, lay down matrix, and restore skeletal continuity while the scaffold disappears as is the case with grafted bone. Unlike autograft bone, such synthetic scaffolds do not have limited availability; there is no risk of disease transfer or rejection as is inherent in unprocessed allograft.

Considerations for engineering synthetic scaffolds involve choosing a material that is biodegradable into biocompatible monomers, that allows cell attachment and proliferation, and that exhibits appropriate mechanical properties. Target properties of replacement material for human trabecular bone are minimum compressive strength of 5 MPa and minimum compressive modulus of 50 MPa (Goldstein et al, 1983). Moreover, the material should be sterilizable and fairly easy to manufacture. Shelf life and availability must also be considered. In addition, potential for drug delivery and cell transplantation as well as easily manipulated physical properties are desired for expanding bone tissue engineering applications.

CURRENT STRATEGIES FOR BONE TISSUE ENGINEERING.

Discussed are four current strategies for using synthetic scaffolds in bone regeneration, each with its own advantages and disadvantages. Common to all four is that the success of each relies on the ability of cells to attach to the scaffold and eventually replace the space formerly occupied by the degradable scaffold. Note that strategies have been categorized and described here essentially for pedagogical purposes; in reality, many bone tissue engineering developments from each strategy may be contained in the design of a single scaffold.

Bone Tissue Induction

Some skeletal defects occur adjacent to relatively healthy bone tissue, such as fracture nonunions where the defect is too large to heal on its own. One approach to filling this type of defect capitalizes on the healthy bone tissue in the neighboring region. In this strategy, a porous degradable scaffold would be used to fill the defect. Such a scaffold would have to be osteoinductive as well as osteoconductive, so that osteoblasts and other cells in the area would be inclined to invade and attach to the scaffold. The cells would grow throughout but not beyond the scaffold: matrix would be laid down and the tissue formed and remodeled as the scaffold degrades. Thus, the area would begin as an unorganized area of porous material and invading tissue but would result in organized bone tissue because of its healing and remodeling potential.

The osteoinductive and osteoconductive capabilities of biomaterials are investigated *in vivo* using two different animal models. To investigate osteoconduction, the material is placed in a defect that would normally heal if left untreated; if normal healing occurs in the presence of the biomaterial, it is considered osteoconductive but not necessarily osteoinductive. To investigate osteoinduction, the material is placed in a defect that will not heal in untreated controls; such a defect may be called a nonunion or critical size defect. If the material can induce bone regeneration in this situation, then it is osteoinductive. Clearly, osteoinduction is a much more difficult criterion to satisfy in scaffold development.

Although this book focuses on polymeric materials, ceramic biomaterials will be briefly discussed for several reasons. Ceramic is another synthetic material being used to make scaffolds for bone regeneration so a description of its merits and drawbacks is helpful for comparison. Also, the minerals that make up the ceramic scaffolds are very similar to the mineral phase of bone tissue. As will be discussed, there is evidence that composites of polymer with particles of these minerals form scaffolds that have enhanced properties over pure polymer or pure ceramic.

Synthetic ceramics are commonly composed of hydroxyapatite (HA), tricalcium phosphate (TCP), or a combination of the two. The material is shaped and sized to fit the defect prior to surgery, that is, prefabricated. HA and TCP degrade very slowly; the total degradation time of TCP may be several years (Altermatt et al, 1992; Bucholz et al, 1987) and still degrades 10 to 20 times faster than HA (Jarcho, 1981). Because of the very low tensile strength of these ceramics, they cannot be used in locations of significant impact, torsion bending, or shear stress. Where they are used, care must be taken to immobilize and shield the area until bone ingrowth has occurred (Gazdag et al, 1995).

Recently another version of the ceramic-related materials has been developed (Constantz et al, 1995). Like bone cement, it is injected, temporarily moldable, and hardens *in situ* at physiologic temperature in about 10 minutes. However, unlike bone cement, it is composed of minerals that could potentially be replaced with natural bone: monocalcium phosphate monohydrate, TCP, and calcium carbonate. This formulation attains a final maximum compressive strength of about 55 MPa and a tensile strength of 2.1 MPa. In one study of implantation in a canine tibial metaphysis the cortical bone region remodeled completely whereas little remodeling occurred in the cancellous bone region. Therefore, this material has potential as an osteoconductive material. In human studies of a percutaneous implantation into an acute fraction of the distal radius, the injectable ceramic provided stabilization during the healing process. This injectable biomaterial holds promise for restoration of normal bone geometry and for fracture stabilization.

Poly(propylene fumarate) (PPF) is being developed as an injectable biodegradable polymeric bone cement to replace the widely used nondegradable PMMA. Briefly, the reaction during defect filling entails the following: unsaturated linear polyester (fluid phase) and filler particles are mixed with a cross-linking agent to form a moldable paste which hardens soon after surgery. N-vinylpyrrolidinone and MMA have been used as cross-linking agents (Gerhart and Hayes, 1989; Sanderson, 1988; Yaszemski et al, 1995). Besides the type of cross-linking agent and filler particles used, many other parameters can be varied such as use of radical initiator, use of accelerator, form of monomeric reactants and reaction scheme. Manipulation of these parameters have been studied to achieve physiologic temperatures and satisfactory handling properties during cross-linking, completion of reaction, reproducibility, and appropriate molecular weight and polydispersity for optimum mechanical strength.

Several formulations of a PPF-based composite cross-linked with N-vinylpyrrolidinone were used in a series of studies; one composite formulation containing β-TCP achieved compressive strength of 4.6 MPa, compressive modulus of 25.8 MPa, and good handling properties. An *in vivo* study of this composite entailed a rat proximal tibia defect (which healed in untreated controls) and demonstrated biocompatibility of the material. During the 5 week study, all defect sites filled with composite showed progressive and extensive bone tissue ingrowth, incorporation of implant into the bone defect, and no evidence of adverse pathologic inflammation (see Figure 1). Bone formation occurred from the periphery of the implant to the center; specifically, replacement of biomaterial in the outermost regions occurred before there was bone in the interior regions (Yaszemski et al, 1995).

The manner in which tissue grows into the PPF suggests the need for a biomaterial that maintains strength *in vivo* over an extended period of time. Two recent studies of PPF-based composites cross-linked by N-vinylpyr-rolidinone (Yaszemski et al, 1996b) and MMA (Frazier et al, 1995) investigated changes in mechanical properties during degradation at physiologic conditions. In both studies, the biomaterials' strength increased for a period

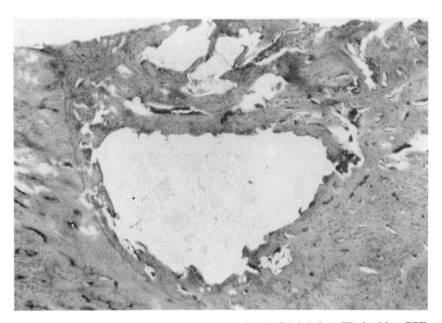

Figure 1. Histological section of a rat cavitational tibial defect filled with a PPF-based composite formulation 4 weeks post-implantation (Yaszemski et al, 1995). Notice absence of any inflammatory response to the material and progressive bone growth from the periphery of the defect to the center. Hematoxylin and eosin stain, 2.5X.

of weeks then decreased. The biomaterial cross-linked with MMA showed a linear increase through day 21 to 4.9 MPa followed by a linear decrease of 1 MPa at day 84. In one formulation of the biomaterial cross-linked with N-vinylpyrrolidinone, the compressive strength measured 8.4 MPa at day 0, 27.1 MPa at day 42, and 17.3 MPa at day 56 (see Figure 2). It is suggested that the increase in mechanical properties during degradation is due to complexing of carboxylic groups by divalent calcium cations or to additional crosslinking (Yaszemski et al, 1996b).

It must be pointed out that none of the studies discussed above showed that the biomaterials used were osteoinductive since none of the models

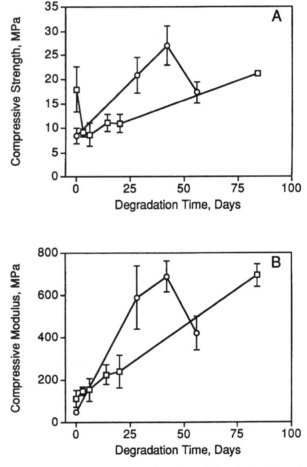

Figure 2. Change in compressive strength and modulus during degradation of two formulations of PPF-based composite (Yaszemski et al, 1996b). Notice that there is an overall increase in mechanical properties during degradation.

used were nonunions. It is possible that osteoinductive scaffolds composed only of materials mentioned thus far (polymer, calcium phosphate salts, etc.) can be engineered purely by altering manufacturing techniques. Indeed, as will be seen it the last section, changes in tissue culture surface alter cellular activity, and therefore even seemingly inert materials may possess bioactivity. However, surface modification studies are still at a relatively preliminary stage so many investigators are turning to other, better characterized systems to induce bone formation: delivery of cells or bioactive molecules on biodegradable scaffolds.

Cell Delivery

Autologous osteoblasts or osteoblast progenitors that are delivered on osteoconductive scaffolds may play an important role in healing nonunion defects. Osteoblast transplantation suggests a head start in bone tissue invasion and extracellular matrix production at the site. Transplanted cells may also release a wide spectrum of growth factors to signal osteoinduction and regeneration. Hence, this strategy may offer many of the advantages of bone grafting while avoiding the complications of immune rejection, donor site morbidity, and limited availability.

Osteoblasts or progenitor cells can be harvested and expanded in culture from bone marrow (Maniatopoulos et al, 1988) or periosteum (Vacanti et al, 1995). The cells can then be seeded on the scaffold and transplanted. *In vivo,* the transplanted cells may proliferate, secrete matrix, and release growth factors as the scaffold vascularizes and slowly degrades. Cellular activity on the material's surface must be extensively characterized when a scaffold for bone regeneration is being developed. Proliferation and migration can be determined by various histology, microscopy, and cell counting techniques used for many adherent cell types. However, response to culture conditions is cell type-specific; therefore it has been efficacious to pinpoint markers for osteoblast function. Widely used assays indicating osteoblast activity include von Kossa staining for mineralization, (^{3}H)-proline incorporation measurements for collagen synthesis, and ALPase activity measurements. Molecular biology techniques are also used to determine production of osteoblast-specific proteins including osteocalcin, osteopontin, and osteonectin.

Culture conditions have been shown to affect osteoblast activity. In particular, recipes of supplements for culture medium have been developed to induce or promote the osteoblast phenotype and function (Aronow et al, 1990; Maniatopoulos et al, 1988). Osteoblasts obtained from fetal calvaria secreted mineralized matrix when L-ascorbic acid and β-glycerolphosphate were added to the medium (Bellows et al, 1990; Nefussi et al, 1985). Dexamethasone, a glucocorticoid, increased the number of collagenous mineralized nodules found in calvaria culture and extended the days that nodules formed in culture (Bellows et al, 1987). Stromal cells obtained from bone marrow also secreted matrix with collagenous, mineralized nodules but only when the medium was supplemented with both β-glycerolphosphate and dexamethasone (Maniatopoulos et al, 1988). Differentiated cultured bone cells provide an essential tool for investigating biodegradable scaffolds for bone tissue regeneration. In addition, enhancement of osteoblast differentiation on scaffolds prior to transplantation may improve effectiveness of this type of bone regeneration therapy.

Poly(α-hydroxy esters) are promising substrates for osteoblast transplantation; three types currently under investigation for this application are poly(lactic acid) (PLA), poly(glycolic acid) (PGA), and poly(lactic-co-glycolic acid) (PLGA). These polymers biodegrade via hydrolysis into biocompatible monomers. Hydrolysis, as opposed to enzymatic degradation, allows a consistent degradation time. Pore sizes, porosity, and degradation rate can be controlled and varied by adjusting processing parameters. In the case of PLGA, the degradation rate can be altered considerably by changing the ratio of lactic to glycolic acid. Such predictability is important in ensuring consistent healing from patient to patient. Finally, some formulations of poly(α-hydroxy esters) have already been approved by the US Food and Drug Association for use in certain surgical procedures.

Osteoblasts cultured on PLGA films were found to attach, proliferate, migrate, and secrete collagen and ALPase at rates comparable to tissue culture polystyrene (Ishaug et al, 1994; 1996b). Long-term three-dimensional *in vitro* studies of porous PLGA foams demonstrated potential for bone regeneration; osteoblasts seeded at a cell density of 6.8×10^5 cells/cm^2 (normalized by area of the top of foam) on scaffolds several millimeters thick demonstrated high ALPase activity, deposited mineralized matrix, and reached a cell density of 1.7×10^6 cells/cm^2 (see Figure 3) (Ishaug et al, 1996a).

In an *in vivo* study, a rat nonunion defect was treated with periosteal cell/polymer construct, chondrocyte/polymer construct, polymer without cells, or nothing (Vacanti et al, 1995). The polymer scaffold was a nonwoven mesh of PGA fibers. At the two time points examined, 6 and 12 weeks, new cartilage formed in the defect filled with chondrocyte/polymer construct, but there was no evidence of new bone formation. The defects filled with nothing or polymer alone showed no gross evidence of bone or cartilage formation at either of the time points. In the defect filled with periosteal cell/polymer complex, there was evidence of bone formation at both time points. Gross and microscopic examination revealed what appeared to be a mixture of tissue types such as cartilage and trabecular bone undergoing bone morphogenesis. This study strongly suggests that polymer constructs delivering bone cells to the defect site have osteoinductive potential.

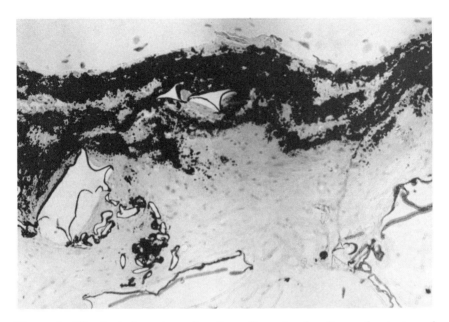

Figure 3. Histological section of a three-dimensional culture of rat stromal osteoblasts seeded on PLGA scaffolds of 90% porosity and 300–500 μm pore size shown 56 days after seeding (Ishaug et al, 1996a). Note significant mineralization stained in black. Von Kossa stain, 40X.

Growth Factor Delivery

Substantial healing may require the incorporation of bioactive molecular signals to prompt events of regeneration. A promising approach is to incorporate various growth factors in the scaffold. Bone morphogenetic proteins (BMPs), first found in demineralized bone matrix (DBM) (Urist, 1965; Urist and Strates, 1971; Urist et al, 1967), are a subclass of the transforming growth factor-beta (TGF-β) superfamily that provide cues for bone development and regeneration (Reddi, 1994). Recently it was shown that demineralized bone matrix exposure significantly increased human osteoblast proliferation and collagen synthesis compared to apatite-based materials or unexposed controls (Zambonin and Grano, 1995). Moreover, systems of BMP incorporation on various scaffolds have been shown to be osteoinductive. Recombinant human BMP-2 (rhBMP-2), delivered on DBM and implanted in 5 mm femoral defects of adult rats, induced new bone formation in a dose-dependent manner (see Figure 4) (Yasko et al, 1992). In untreated controls, this critical rat femoral defect resulted in a nonunion. Collagen carriers have been used to deliver BMPs to regenerated bone in a variety of animal studies such as rat craniotomies (Marden et al, 1994), mandibular recon-

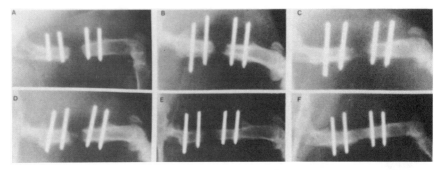

Figure 4. Radiographs of healing of long bone segmental defects in rats using low or high doses of rhBMP-2 (1.4 or 11μg): two untreated controls at 4.5 weeks (A and B), low-dose rhBMP-2 at 4.5 and 6 weeks (C and D), and high-dose rhBMP-2 at 4.5 and 6 weeks (E and F). The high-dose specimens shown achieved 75% healing at 4.5 weeks and union of defect at 6 weeks. (Reproduced with permission from Yasko et al, 1992.)

struction (Toriumi et al, 1991), and large segmental bone defects (Cook et al, 1994); however, collagen may provoke an immune response. Proposed is to develop a modified poly(α-hydroxy ester) that mimics collagen's ability to deliver BMPs at skeletal defect sites and to promote osteoconduction. Such a system should express pharmacokinetics of BMP-release that match the local needs of the bone wound continuum and should position BMPs for spacially favorable sites for cell interaction (Hollinger and Leong, 1996).

TGF-β1 seems to have varying effects on culture systems depending on cell type and dose. However, there is little doubt that it is involved in bone formation, fracture healing, and remodeling (Liebergall et al, 1994). In a study using a nonunion model, cells were incubated in TGF-β1 then delivered on ceramic to the defect site. TGF-β1 improved bone regeneration compared to devices in which cells were not incubated with TGF-β1 (Liebergall et al, 1994). This growth factor also increased bone ingrowth in another *in vivo* study; titanium rods for canine bilateral implantation were plasma-flame-sprayed with HA and TCP (Sumner et al, 1995). The implants treated with TGF-β1 showed a 3-fold increase in bone ingrowth over those that were not.

Bone Flaps

Bone growth into scaffolds will not occur in some defects created by bone tumor removal or in other hostile environments where extensive tissue damage has occurred. The *in vitro* development of polymer-cell constructs large enough to fill critical size defects is hindered by the formidable obstacle of nutrient diffusion limitations to the innermost portion of the construct.

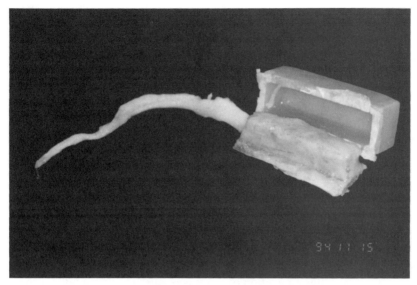

Figure 5. Photograph of a vascularized bone flap with a single vascular peticle created in sheep after 6 weeks by implanting $4 \times 1 \times 1$ cm^3 chambers filled with morcellized bone graft adjacent to rib periosteum (Miller et al, 1996).

An ingenious strategy to forego the osteoinduction requirement and to solve the upscale dilemma is to prefabricate bone flaps *in vivo*, that is, to induce ectopic formation of bone tissue in a non-hostile location of the patient's body where it can prevascularize. This autologous neo-osseous tissue can then be harvested and transplanted to the defect site where blood vessels can be microsurgically attached to existing vessels. Such vascularized bone flaps have been successfully created after 6 weeks by filling 0.5 to 1.0 cm thick chambers with morcellized corticocancellous bone and implanting them adjacent to the sheep rib periosteum (see Figure 5) (Miller et al, 1996). The new bone assumed the shape of the chambers employed. In an effort to avoid donor site morbidity, synthetic materials, such as PLGA, are being investigated to fill the chamber. Such materials would have to fulfill the same requirements as previously mentioned scaffolds, but they would have the advantage of being placed in a highly vascularized environment.

MANUFACTURING GOALS

In the various strategies described above, it is clear that specific objectives must dictate design. Researchers have developed a wide variety of manufacturing techniques to meet therapeutic objectives. Porosity and pore sizes have been characterized and varied to allow optimum cell invasion. Combining polymer with various particulates may increase strength or osteoinductivity. Mechanical properties of scaffolds during degradation and bone regeneration must be characterized and sufficient for expected loads. Depending on application, scaffolds may be prefabricated for subsequent transplantation or injectable for *in situ* molding. Finally, there is increasing evidence that changes in scaffold surface chemistry and topography alter cellular activity. Therefore, surfaces may need to be characterized or even altered to facilitate bone tissue regeneration.

Filler Particulates

Pores are routinely created in scaffolds to allow and/or promote three-dimensional tissue growth, nutrient diffusion, and vascularization. They are created by incorporating a porogen such as salt or gelatin particles of a specific size prior to molding. After hardening, the porogen is removed by leaching. Ample interconnection of pores is required for complete leaching. When this requisite has been fulfilled, pore size and porosity approximate initial particle size and particle weight fraction, respectively (see Figure 6) (Thomson et al, 1995b).

The size of scaffold pores must be large enough to allow the circumferential attachment of cells, yet small enough to encourage migration and proliferation. Osteoblasts cultured in calcium phosphate ceramic prefer a pore size of 200 mm (Dennis et al, 1992). It has been proposed that this pore size possesses a curvature that optimizes the compression and tension of the cell's mechanoreceptors (Boyan et al, 1996). Indeed, studies have revealed altered gene expression and proliferation in osteoblasts subjected to mechanical stimuli (Copley et al, 1994; Dolce et al, 1995). However, there is concern that optimal pore size in ceramics may not generalize for all biomaterials (Robinson et al, 1995). For example, when PLA was implanted in calvarial rat defects, pore sizes of 300–350 mm supported bone ingrowth while smaller sizes did not (Robinson et al, 1995). Yet in another study, osteoblasts showed no significant difference in proliferation or function when seeded on PLGA foams with pore sizes of either 150–300 μm, 300–500 μm, or 500–710 μm (Ishaug et al, 1996a). The importance of determining optimum values for the specific system being used cannot be underestimated and will be illustrated again in this chapter.

Composites of two or more complementary materials may greatly enhance success of various scaffolds, and several formulations are currently being studied for bone tissue engineering. PLGA possesses many favorable characteristics, but its compressive strength is below design goals. By combining the PLGA with 30% hydroxyapatite fibers, 35% porous PLGA shows a 3-fold increase in compressive strength and a 2-fold increase in

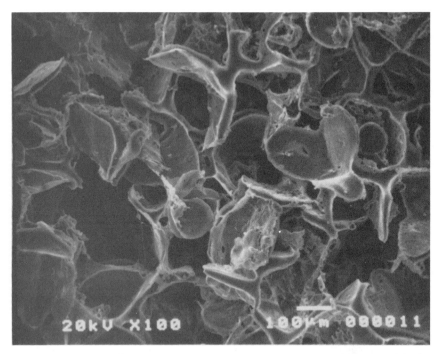

Figure 6. SEM photomicrograph of PLGA porous scaffolds. Pores created by incorporating 90 weight percent salt particles with a size range of 150–300 μm (Ishaug et al, 1996a).

compressive modulus (Thomson et al, 1995a). Hydroxyapatite may also enhance cell differentiation; two types of apatite-based biomaterials prevented human osteoblast proliferation but increased collagen synthesis with respect to controls not exposed to biomaterial (Zambonin and Grano, 1995). In the development of a biodegradable bone cement, β-TCP has been combined with PPF to serve as an osteoconduit upon which new bone can grow (Yaszemski et al, 1995). Development of new production techniques and thorough characterization of polymers, ceramics, and their composites will continue to play a vital role in improving bone regeneration therapies in the future.

Prefabrication versus In Situ Molding

Prefabricated scaffolds are manufactured prior to surgery in the size and shape to fill the defects. Preparation of prefabricated PLGA scaffolds has been well-studied and relatively reproducible; the procedure has been fine-tuned and repeated in several studies (Ishaug et al, 1996a; Thomson et al,

1995ab). Moreover, PLGA and its degradation products have been shown to be biocompatible. Hence, attention has turned to modifying the polymer to improve osteoinduction and mechanical properties. Examples are incorporation of growth factors or strengthening fibers, respectively, as discussed above.

Injectable biomaterials, on the other hand, must meet additional requirements since human tissue is potentially exposed to reactants, products, and reaction temperatures. At the beginning of surgery, polymeric injectables are still in monomeric form. Polymerizing or cross-linking agents are added at the onset of the reaction, possibly along with a radical initiator and an accelerator. It is clear that any monomer or agent that may come in contact with tissue prior to completion of the reaction must be biocompatible. In addition, reactions must occur near physiologic temperatures and material must be easy to handle. Finally, to ensure total compliance with these requirements, the reaction must be easily reproducible. Although this is no easy task, there has been some success with perfecting this process in ceramics and PPF-based materials, as previously discussed.

Surface Modification: An Increasingly Important Consideration

The biomaterial surface plays a prominent role in cellular activity. As interactions between cells and the extracellular matrix become better understood, investigators are finding increased success in manipulating cellular activity via surface topology and surface bioactivity. Mechanical and chemical stimuli, too, can have appreciable impact on function and proliferation.

Due to differences in charge density, net polarity of charge, and surface chemistry, osteoblast response varies with the material on which cells are cultured (Boyan et al, 1996). Some of these variations have been attributed to proteins present in the medium that adsorb onto the surface to different degrees or with different structural arrangements (Norde, 1992). Surface texture is another important characteristic; a higher percentage of osteoblastlike cells cultured on commercially pure Ti attached to rougher surfaces than to smooth surfaces (Bowers et al, 1992). Another study using the same material showed highest osteocalcin content and ALPase activity on smooth, polished surfaces compared to rough surfaces (Stanford et al, 1994).

Patterned tissue culture surfaces have also been manufactured for investigating surface chemistry-specific attachment and function (see Figure 7). Recently, a substrate was prepared which expressed parallel line patterns of 100 μm-wide regions of dimethyldichlorosilane (DMS) alternated with 50 μm-wide regions of N-(2-aminoethyl)-3-aminopropyl-trimethoxysilane (EDS) or quartz (Healy et al, 1996). In a serum-enriched medium, osteoblasts adhered preferentially to EDS or quartz instead of DMS (see Figure 8). Since serum proteins adsorbed onto both EDS and DMS, it is

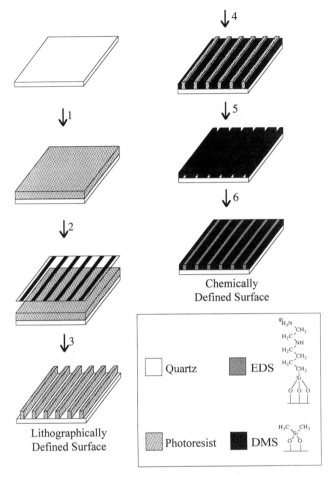

Figure 7. Patterning technique for immobilizing alkyloxysilanes on quartz substrates. (Patterning of EDS is an analogous process.) (1) Quartz is coated with photoresist. (2) Photoresist is exposed to UV light through a photomask. (3) UV-exposed bands are developed away. (4) DMS is bound to quartz. (5) Remaining photoresist is removed. (6) Amine is bound to quartz. (Reproduced with permission from Healy et al, 1995.)

postulated that protein molecules in the serum adsorbed onto EDS or on quartz in the correct composition or conformation (Healy et al, 1996). Another possibility is that the protein molecules adsorbing onto EDS are different than those adsorbing onto DMS (Healy et al, 1996).

Cell-surface interactions are highly complex, and further investigation is required to characterize phenomena and elucidate trends. Already it is clear, though, that cellular response may be affected by material type, material

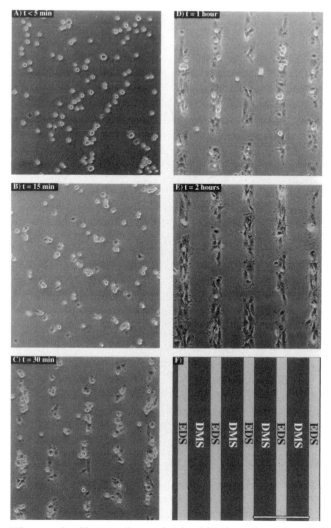

Figure 8. Micrograph of bone cell organization kinetics using a time-lapse micros-copy technique. Note migration of cells from DMS region to EDS region. Cells were plated using 15% fetal bovine serum. (Reproduced with permission from Healy et al, 1995.)

surface, protein adsorption, cell type, and cell maturation (Boyan et al, 1996; Healy et al, 1996). As more information is uncovered, scaffolds for bone regeneration can be modified to optimize osteoblast activity.

A specific modification for scaffolds already being investigated is the attachment of bioactive peptide sequences to the substrate. Certain peptide sequences, such as Arg-Gly-Asp-Ser (RGDS) occur in many ECM components and interact with integrin cell membrane receptors at cell adhesion focal points (Yamada, 1991). ECM components of bone tissue with RGD include collagen, fibronectin, and osteopontin (Gehron et al, 1993). In one study, osteoblasts were cultured on glass modified with the RGD peptide or nonadhesive, scrambled sequence and in the presence or absence of BMP-7 (Dee et al, 1996). The culture with a combination of RGD substrate and BMP-7 showed a substantial increase in mineralization in 21 days over all other combinations of treatments. Because of its role in both attachment and differentiation, RGD incorporation may contribute greatly to scaffold osteoinductivity and bone regeneration.

CONCLUSION

At each step in the development of a biodegradable scaffold for bone regeneration, it is important to answer certain critical questions. Are the materials to which tissue is exposed biocompatible? Does the material have satisfactory properties? Will the cells function on the material as required for bone growth? If not, what bioactive molecules must be present and how will they be delivered? How do cellular function and material properties change when a different manufacturing technique or material formulation is used? Already much progress has been made in scaffold development. Degradation rates of polymers can be controlled by changing formulation and processing parameters. Filler particulates may be used to make pores or vary mechanical properties. Delivery of bone cells or growth factors seem to be critical in healing nonunion fractures; seeding density or dose, respectively, are currently being optimized. Development is particularly challenging, though, because so many parameters affect the success of a biodegradable scaffold. With thoughtful, thorough, and reproducible investigation, these parameters are becoming better understood.

References

Altermatt S, Schwobel M, Pochon JP (1992): Operative treatment of solitary bone cysts with tricalcium phosphate ceramic: A 1 to 7 year follow-up. *Eur J Pediatr Surg* 2:180–182

Aronow MA, Gerstenfeld LC, Owen TA, Tassinari MS, Stein GS, Lian JB (1990): Factors that promote progressive development of the osteoblast phenotype in cultured fetal rat calvaria cells. *J Cell Phys* 143:213–221

Bellows CG, Aubin JE, Heersche JN (1987): Physiological concentrations of glucocorticoids stimulate formation of bone nodules from isolated rat calvaria cells in vitro. *Endocrinology* 121:1985–1992

Bellows CG, Heersche JN, Aubin JE (1990): Determination of the capacity for proliferation and differentiation of osteoprogenitor cells in the presence and absence of dexamethasone. *Dev Biol* 140:132–138

Bobyn JD, Mortimer ES, Glassman AH, Engh CA, Miller JF, Brooks CE (1992): Producing and avoiding stress shielding. Laboratory and clinical observations of noncemented total hip arthroplasty. *Orthop Rel Res* 274:79–96

Bowers KT, Keller JC, Randolph BA, Wick DG, Michaels CM (1992): Optimization of surface micromorphology for enhanced osteoblast responses *in vitro*. *Int J Oral Maxillofac Imp* 7:302–310

Boyan BD, Hummert TW, Dean DD, Schwartz Z (1996): Role of material surfaces in regulating bone and cartilage cell response. *Biomaterials* 17:137–146

Bucholz RW, Carlton A, Holmes RE (1987): Hydroxyapatite and tricalcium phosphate bone graft substitutes. *Orthop Clin North Am* 18:323–334

Buckwalter JA, Glimcher MJ, Cooper RR, Recker R (1995a): Bone biology part I: Structure, blood supply, cells, matrix, and mineralization. *J Bone Joint Surg* 77-A:1256–1275

Buckwalter JA, Glimcher MJ, Cooper RR, Recker R (1995b): Bone biology part II: Formation, form, modeling, remodeling, and regulation of cell function. *J Bone Joint Surg* 77-A:1276–1289

Constantz BR, Ison IC, Fulmer MT, Poser RD, Smith ST, VanWagoner M, Ross J, Goldstein SA, Jupiter JB, Rosenthal DI (1995): Skeletal repair by *in situ* formation of the mineral phase of bone. *Science* 267:1796–1799

Cook SC, Baffes GC, Wolfe MW, Sampath K, Rueger DC, Whitecloud TS (1994): The effect of human recombinant osteogenic protein-1 on healing of large segmental bone defects. *J Bone Joint Surg* 76A:827–838

Copley LA, Reilly TM, Brighton CT (1994): Integrins and the transduction of mechanical stress into proliferation in rat osteoblasts. *Trans Orthop Res Soc* 19:306

Dee KC, Rueger DC, Anderson TT, Bizios R (1996): Conditions which promote mineralization at the bone-implant interface: A model *in vitro* study. *Biomaterials* 17:209–215

Dennis JE, Haynesworth SE, Young RG, Caplan AI (1992): Osteogenesis in marrow-derived mesenchymal cell porous ceramic composites transplanted subcutaneously: Effect of fibronectin and laminin on cell retention and rate of osteogenic expression. *Cell Transpl* 1:23–32

Dolce C, Kinniburgh AJ, Dziak R (1995): Proto-oncogene activation in osteoblast cells due to mechanical stretching. *J Dent Res* 74:153

Frazier DD, Lathi VK, Gerhart TN, Altobelli DE, Hayes WC (1995): *In-vivo* degradation of a poly(propylene-fumarate) biodegradable particulate composite bone cement. *Mater Res Soc Symp Proc* 394:15–19

Gazdag AR, Lane JM, Glaser D, Forster RA (1995): Alternatives to autogenous bone graft: Efficacy and indications. *J Am Acad Orthop Surg* 3:1–8

Gehron RP, Fedarko NS, Hefferan TE (1993): Structure and molecular regulation of bone matrix proteins. *J Bone Min Res* 8 (suppl 2):S457–S65

Gerhart TN, Hayes WC (1989): Bioerodible implant composition. *United States Patent* 4,843,112:1–16

Goldstein SA, Wilson DL, Sonstegard DA, Matthews LS (1983): The mechanical properties of human tibial trabecular bone as a function of metaphyseal location. *J Biomech* 16:965–976

Healy KE, Thomas CH, Rezania A, Kim JE, McKeown PJ, Lom B, Hockberger PE

(1996): Kinetics of bone cell organization and mineralization on materials with patterned surface chemistry. *Biomaterials* 17:195–208

Hollinger JO, Leong K (1996): Poly(α-hydroxy acids): Carriers for bone morphogenetic proteins. *Biomaterials* 17:187–194

Ishaug SL, Yaszemski MJ, Bizios R, Mikos AG (1994): Osteoblast function on synthetic biodegradable polymers. *J Biomed Mater Res* 28:1445–1453

Ishaug SL, Crane GM, Miller MJ, Yaszemski MJ, Mikos AG (1996a): Bone formation by three-dimensional stromal osteoblast culture in biodegradable polymer scaffolds. *J Biomed Mater Res:* in press

Ishaug SL, Payne RG, Yaszemski MJ, Aufdemorte TB, Bizios R, Mikos AG (1996b): Osteoblast migration on poly(α-hydroxy esters). *Biotechn Bioeng:* 50:443–451

Jarcho M (1981): Calcium phosphate ceramics as hard tissue prosthetics. *Clin Orthop* 157:259–278

Liebergall M, Young RG, Ozawa N, Reese JH, Davy DT, Goldberg VM, Caplan AI (1994): The effects of cellular manipulation and TGF-β in a composite bone graft. In: *Bone Formation and Repair,* Brighton CT, Friedlaender GE, Lane JM, eds. Rosemont, IL: American Academy of Orthopaedic Surgeons

Maniatopoulos C, Sodek J, Melcher AH (1988): Bone formation *in vitro* by stromal cells obtained from bone marrow of young adult rats. *Cell Tissue Res* 254:317–330

Marden LJ, Hollinger JO, Chaudhari A, Turek T, Schaub R, Ron E (1994): Recombinant bone morphogenetic protein-2 is superior to demineralized bone matrix in repairing craniotomies defects in rat. *J Biomed Mater Res* 28:1127–1138

Miller MJ, Goldberg DP, Yasko AW, Lemon JC, Satterfield WC, Wake MC, Mikos AG (1996): Guided bone growth in sheep: a model for tissue-engineered bone flaps. *Tissue Eng:* 2:51–59

Nefussi JR, Boy-Lefevre ML, Boulekbache H, Forest N (1985): Mineralization *in vitro* of matrix formed by osteoblasts isolated by collagenase digestion. *Differentiation* 29:160–168

Norde W (1992): The behavior of proteins at interfaces, with special attention to the role of the structure stability of the protein molecule. *Clin Mater* 11:85–91

Petty W, Spanier S, Shuster JJ (1988): Prevention of infection after total joint replacement. *J Bone Joint Surg* 70–A:536–9

Reddi AH (1994): Bone and cartilage differentiation. *Curr Opin Genet Dev* 4:737–744

Robinson B, Hollinger JO, Szachowicz E, Brekke J (1995): Calvarial bone repair with porous D,L-polylactide. *Otolaryngol Head Neck Surg* 112:707–713

Sanderson JE (1988): Bone replacement and repair putty material from unsaturated polyester resin and vinyl pyrrolidone. *United States Patent* 4,722,948:1–14

Stanford CM, Keller JC, Solursh M (1994): Bone cell expression on titanium surfaces is altered by sterilization treatments. *J Dent Res* 73:1061–1071

Sumner DR, Turner TM, Purchio AF, Gombotz WR, Urban RM, Galante JO (1995): Enhancement of bone ingrowth by transforming growth factor-β. *J Bone Joint Surg* 77-A:1135–1147

Thomson RC, Yaszemski MJ, Powers JM, Harrigan TP, Mikos AG (1995a): Poly(α-hydroxy ester)/short fiber hydroxyapatite composite foams for orthopedic application. *Mater Res Soc Symp Proc* 394:25–30

Thomson RC, Yaszemski MJ, Powers JM, Mikos AG (1995b): Fabrication of biodegradable polymer scaffolds to engineer trabecular bone. *J Biomater Sci Polym Edn* 7:23–38

Toriumi DM, Kotler HS, Luxunberg DP, Holtrop ME, Wang EA (1991): Mandibular reconstruction with a recombinant bone-inducing factor: Functional, histologic, and biomechanical evaluation. *Arch Otolaryngol Head Neck Surg* 117:1101–1112

Urist MR (1965): Bone: Formation by autoinduction. *Science* 150:893–899

Urist MR, Strates BS (1971): Bone morphogenetic protein. *J Dent Res* 50:1392–1406

Urist MR, Silverman MF, Buring K, Dubuc FL, Rosenburg JM (1967): The bone induction principle. *Clin Orthop Rel Res* 53:243

Vacanti CA, Kim W, Upton J, Mooney D, Vacanti JP (1995): The efficacy of periosteal cells compared to chondrocytes in the tissue engineered repair of bone defects. *Tissue Eng* 1:301–308

Yamada KM (1991): Adhesive recognition sequences. *J Biol Chem* 266:12809–12812

Yasko AW, Lane JM, Fellinger EJ, Rosen V, Wozney JM, Wang EA (1992): The healing of segmental bone defects, induced by recombinant human bone morphogenetic protein (rhBMP-2). *J Bone Joint Surg* 74-A:659–670

Yaszemski MJ, Payne RG, Hayes WC, Langer RS, Aufdemorte TB, Mikos AG (1995): The ingrowth of new bone tissue and initial mechanical properties of a degrading polymeric composite scaffold. *Tissue Eng* 1:41–52

Yaszemski MJ, Payne RG, Hayes WC, Langer RS, Mikos AG (1996a): Evolution of bone transplantation: Molecular, cellular, and tissue strategies to engineer human bone. *Biomaterials* 17:175–185

Yaszemski MJ, Payne RG, Hayes WC, Langer RS, Mikos AG (1996b): The *in vitro* degradation of a poly(propylene fumarate)-based composite material. *Biomaterials:* 17:2127–2130

Zambonin G, Grano M (1995): Biomaterials in orthopedic surgery: Effects of different hydroxyapatites and demineralized bone matrix on proliferation rate and bone matrix synthesis by human osteoblasts. *Biomaterials* 16:397–402

13

TISSUE ENGINEERING OF THE LIVER

HANMIN LEE, M.D. AND JOSEPH P. VACANTI, M.D.

INTRODUCTION

Approximately 30,000 people a year in the United States alone die from end-stage liver disease, and many others suffer debilitating illness (American Liver Foundation, 1988). The only accepted current treatment is liver transplantation, which, while effective, has many drawbacks, including the need for life-long immunosuppression, the possibility of rejection, and long hospital stays. The most significant problem with liver transplantation is the lack of donor organs.

In 1994 only 3652 liver transplants were performed in the United States due to lack of available donor organs (US Scientific Registry for Organ Transplantation, 1990). Many patients needing donor livers die on the waiting list, while many others suffer debilitating complications of end-stage liver disease waiting for a donor liver.

As time spent on the waiting list is continuously increasing and wait lists are becoming progressively longer, therapeutic alternatives to liver transplantation are being explored. One way to increase available treatment to patients with liver disease is to increase the quantity of donor organs. Some groups, particularly in Europe, are using split-liver grafts to expand the donor pool (Emond et al, 1990): In this technique, the left lobe of the donor organ is transplanted to one recipient while the right lobe is transplanted to another. Others are offering living-related liver transplantation as a way of increasing donor organs (Broelsch et al, 1994). This is of particular importance in countries such as Japan where cadaveric organ transplantation is not performed. Both of these methods are currently in clinical trials and have shown promising initial results. The liver

Synthetic Biodegradable Polymer Scaffolds
Anthony Atala and David Mooney, Editors
Robert Langer & Joseph P. Vacanti, Associate Editors
© 1997 Birkhäuser Boston

transplant group in Pittsburgh has attempted a xenogenic liver transplantation using a baboon liver for a human recipient (Starzl et al, 1993). All of these methods, while increasing the donor pool of available organs, would still necessitate that the patient take life-long immunosuppressants with their attendant side-effects.

Other experimental solutions include the use of extracorporeal liver support systems (Yarmush et al, 1992). Most of these systems use xenogenic hepatocytes or human hepatoblastoma cell lines in an extracorporeal bioreactor that is then exposed to a patient's blood through a vascular conduit. The hepatocytes are able to replace some liver function. These extracoporeal liver support systems have been shown to have some success in treating acute liver failure, particularly as a bridge to transplantation while awaiting a donor liver (Rozga et al 1994; Sussman et al, 1992).

Still other groups are experimenting with hepatocellular transplantation, either orthopically or heterotopically. Many groups have shown successful orthotopic transplantation of hepatocytes when injected either into the spleen or into the portal vein in experimental animals and human trials (Grossman et al, 1994; Matas et al, 1976; Onoder et al, 1992; Ponder et al, 1991). The injected hepatocytes are carried from the spleen or portal vein and have been reported to engraft into the native liver. Others have injected hepatocytes into the peritoneal cavity either as a cell suspension or on microcarriers. Our approach towards a potential therapy for hepatic replacement has been heterotopic transplantation of hepatocytes using three-dimensional polymer matrices as cell delivery devices (Langer and Vacanti, 1993). In this chapter, we will focus on approaches to replacement of liver function that can be broadly categorized as "tissue engineering", including extracorporeal liver support systems and hepatocellular transplantation on microcarrier devices or on synthetic polymers.

Extracorporeal Liver Support Systems

Several groups have developed extracorporeal liver assist devices to temporarily replace liver function that would function similarly to renal dialysis machines for kidney failure or to extracorporeal membrane oxygenators (ECMO) for lung failure. Unlike the renal dialysis machine or the ECMO machines which are entirely synthetic, these liver assist devices are bioartificial. The device consists of a circuit that pumps blood away from the patient into a bioreactor that houses either hepatocytes or hepatoblastoma cells which retain differentiated hepatic function. The blood is exposed to the cells which secrete necessary proteins into the blood and metabolize toxins from the blood. The blood is then pumped back into the patient's circulatory system.

Rozga et al reported improvements in seven patients with severe acute liver failure who were treated with an extracorporeal liver assist device using porcine hepatocytes (Rozga et al, 1994). A hollow fiber bioreactor housed approximately 4–6 × 10^9 hepatocytes attached to collagen-coated dextran microcarrier beads. The bioreactor was designed with a large internal surface area (6000cm^2) in order to maximize the hepatocyte-plasma interface. Patients showed clinical signs of improvement including better neurologic status and decreased plasma ammonia. All patients survived and ultimately received a liver transplant. Sussman et al have used tumor derived hepatoblastoma cells in a similar system (Sussman et al, 1992).

Gerlach et al have reported a more complex liver assist device with four separate "capillary" compartments (Gerlach et al, 1995). These capillary compartments are composed of four separate networks of hollow microfibers. Two capillary compartments were designed for inflow and outflow. The inflow capillary network was constructed of polyamide while the outflow was constructed of hydrophilic polypropylene. A third capillary compartment composed of hydrophobic polypropylene or silicone was used for oxygen delivery. The fourth compartment was composed of matrigel-coated hydrophilic propylene and was seeded with a co-culture of sinusoidal endothelial cells and hepatocytes. The four compartments are intimately interwoven to allow for substrate, metabolite, and gas exchange amongst all four compartments.

The extracorporeal liver assist devices appear to be a promising temporizing treatment for end-stage liver disease. They could serve either as a bridge to transplantation in cases of irreversible hepatic damage or as a bridge to recovery in instances of reversible acute fulminant hepatitis.

HEPATOCELLULAR TRANSPLANTATION ON MICROCARRIERS

One vehicle for delivery of hepatocytes for transplantation is microcarrier beads. Hepatocytes are seeded onto microcarrier beads which are then injected into an implantation site. Demetrious et al described the transplantation of hepatocytes on type I collagen-coated dextran microcarrier beads. The cells were seeded onto the microbeads and then implanted intraperitoneally. Hepatocyte specific function of the transplanted cells was assessed using two enzyme-deficiency rat models. The Gunn rats have a deficiency in UDP-glucuronlytransferase and are unable to conjugate bilirubin, resulting in an elevation of unconjugated bilirubin and the absence of conjugated bilirubin in the serum. The Nagase rats have a defect in albumin synthesis and have very low serum levels of albumin. Normal hepatocytes on microcarrier beads implanted into the Gunn and Nagase rats resulted in improvements in levels of serum conjugated bilirubin and albumin, respectively (Demetriou et al, 1986).

HEPATOCELLULAR TRANSPLANTATION ON POLYMERIC SCAFFOLDS

Over the past decade, our laboratory has focused on hepatic replacement using synthetic, degradable, polymeric scaffolds as delivery devices for hepatocytes. Hepatocytes are isolated using collagenase digestion and are then seeded onto the polymer. The hepatocytes are allowed to attach to the polymer for a variable length of time and then the polymer/hepatocyte construct is implanted into the animal. The potential benefits of this technique include the ability to transplant a large number of cells, induction of neovascularization, and the ability to give structural cues to the hepatocytes as they proliferate and reorganize.

Hepatocyte cell isolation

Hepatocytes are isolated from experimental animals using a standard two-step collagenase digestion technique as described by Seglen and modified by Aiken et al (Aiken et al, 1990; Seglen, 1976). Following an abdominal incision, the inferior vena cava below the liver is cannulated, and the inferior vena cava above the liver is ligated to allow perfusion retrograde through the liver via the hepatic veins. The animal is systemically anticoagulated with heparin to prevent thrombus formation in the liver and to allow even perfusion. The portal vein is incised to allow drainage of the effluent. The liver is perfused with an isotonic, buffered saline solution to remove the blood within the liver. Next, the liver is perfused with an isotonic, buffered solution containing calcium and collagenase to digest the extracellular matrix of the liver. After the liver has been digested, it is excised and placed in media. The liver is then mechanically agitated to disperse the hepatocytes into a single cell suspension. The nonparenchymal cell fraction, consisting of sinusoidal endothelial cells, Kuppfer cells, Ito cells, and biliary epithelial cells can be separated from the hepatocyte fraction using a low speed centrifugation. Each of the subpopulations of nonparenchymal cells can then be purified using a variety of techniques. Viable hepatocytes are then purified using a percoll gradient centrifugation step.

Cell number and viability are determined using trypan blue exclusion. The cells are then seeded onto the polymers. We have found 5×10^7 hepatocytes/ml as a concentration that saturates the polymer with hepatocytes.

Hepatocyte/polymer implantation and retrieval

The hepatocyte/polymer construct is then implanted in a highly vascularized site. The mesentery has proven to be an ideal site, as it is highly vascularized and has a large surface area for implantation of constructs (Johnson, 1994). The construct is placed within leaves of the mesentery and surgically sewn in place. The constructs are harvested after a variable length of time *in vivo* for analysis.

Analysis of results

Specimens are chemically fixed and paraffin embedded for histologic and immunohistochemical analysis. Hematoxylin and Eosin (H&E) staining of specimens have demonstrated viable hepatocytes within the constructs. Hepatocytes are easily identified by the large size of the cell and nucleus, and columnar shape. Immunohistochemical staining with albumin has shown hepatocyte-specific function. Using computer-assisted morphometric analysis of H&E sections, the area of engrafted hepatocytes within the polymers can be calculated, allowing for comparison of various hepatotrophic stimuli. Serologic markers for enzyme function in replacement models have also been tested and will be described below.

Polymers

Polymers used for hepatocyte transplantation in our laboratory have been poly-glycolic acid (PGA) fibers, poly-lactic acid (PLA) foams, copolymers of glycolic and lactic acids (PGLA), poly-vinyl alcohol (PVA) foams, polyorthoesters, and polyanhydride. The PGA fiber sheets (Albany International, Albany, NY) and PVA foam sponges (Unipoint Industries, Highpoint, NC) are available from industrial sources for research purposes. Polyglactin is a 90-10 copolymer of glycolide and lactide acid that is produced as Vicryl suture material (Ethicon, Somerville, NJ). PLA foams are manufactured by dissolving PLA in chloroform and packing this solution into Teflon cylinders packed with sodium chloride particles. The chloroform is allowed to evaporate, and the salt particles are removed by leaching the polymers in distilled water. This technique yields polymeric foams of PLA whose porosity can be manipulated by varying the size of the salt particles and varying the salt/polymer ratio. Porosity of 95% has been achieved (Mooney et al, 1994). Polyanhydride and polyorthoester discs or filaments are fabricated using either solvent casting, compression molding, or filament drawing techniques (Vacanti et al, 1988.).

In *in vitro* studies, hepatocytes were shown to attach to the polymer and survive in culture (Mooney et al, 1992; Vacanti et al, 1988). In the initial studies described by Vacanti et al, rodent hepatocytes were seeded onto polyglactin, polyanhydride, and polyorthoester polymer scaffolds. Hepatocytes were shown to attach to all three of the above polymers. Cell viability was assessed after 3–4 days of the polymer/hepatocyte construct in culture. Hepatocytes on polyglactin proved to have the highest viability. Cell-polymer constructs implanted *in vivo* showed successful engraftment of small groups of hepatocytes seven days after implantation (Vacanti et al, 1988).

Based on the success of the polyglactin polymers, our laboratory used PGA meshes for much of our initial experiments in hepatocyte transplantation. PGA meshes are composed of fibers that are approximately 15μ in diameter with pore sizes of 250μ and up to 99% porosity. The filamentous pattern of the PGA meshes provides a large area for cell attachment as well as large pores for diffusion of nutrients and waste.

PVA discs were first used to improve quantification of transplanted cell mass. The PVA foams can be manufactured in a consistent shape and size and maintain their form after implantation over time. Thus, consistent data from morphometric analysis of hepatocyte cell area could be obtained. This allowed for a reproducible way to compare the effect of different stimuli on the survival of hepatocytes within the constructs. However, the PVA foams are not biodegradable and thus would not be considered for clinical use as they would lead to a chronic inflammatory response.

PLA discs can be manufactured to a specific size and shape and are biodegradable. They maintain a definable shape for at least 8 weeks after implantation and thus can be analyzed with computer-assisted morphometric analysis like the PVA discs. Additionally, as they are biodegradable, they leave no permanent foreign body reaction (Mooney et al, 1994) (Figure 1).

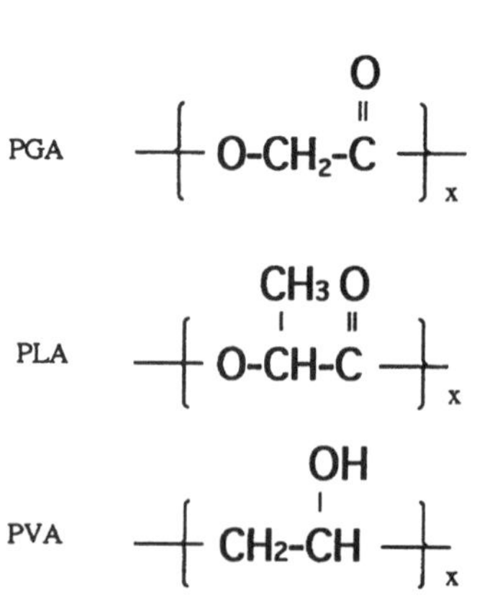

Figure 1. Chemical backbone of Polyglycolic Acid (PGA), Polylactic Acid (PLA), and Polyvinyl alcohol (PVA), commonly used polymers for tissue engineering of liver.

Hepatotrophic stimulation

While initial results showed survival of hepatocytes on PGA polymers after transplantation, the engraftment was relatively limited. In order to increase engraftment of the hepatocytes, our laboratory has investigated several methods to stimulate the proliferation of transplanted hepatocytes. Two well-known ways to create a hepatotrophic environment are partial hepatectomy and portacaval shunt.

It has long been known that the portal blood has growth factors that are important for hepatocyte growth and survival (Bucher, 1991). Some of these factors are known, such as hepatocyte growth factor (HGF), epidermal growth factor (EGF), insulin, and glucagon. When the portal blood is diverted from the liver by creation of a portacaval shunt (PCS), the liver loses direct stimulation of the hepatotrophic substances and becomes atrophic. The PCS is created by dividing the portal vein and attaching it end-to-side to the inferior vena cava. The portal blood is then distributed systemically without first-pass metabolism by the liver (Figure 2).

Theoretically, the systemic circulation should then be enriched in hepatotrophic factors. PCS in rats one to two weeks prior to implantation of hepa-

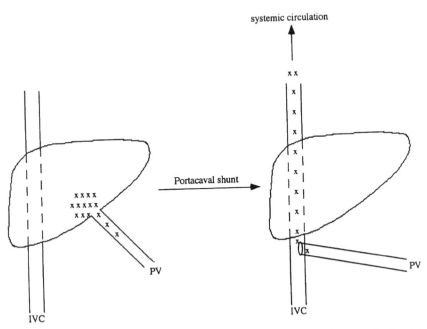

Figure 2. Schematic diagram of a portacaval shunt. Normally, the portal vein, rich with hepatotrophic factors (x) supplies blood to the liver. After portacaval shunt, the portal blood is distributed to the systemic circulation, bypassing first-pass metabolism by the liver.

tocyte-polymer constructs has been shown to increase engraftment of hepatocytes by a six to thirtyfold increase of engrafted hepatocytes in specimens harvested one to four weeks after implantation according to morphometric analysis (Cusick et al, 1995; Kaufmann et al, 1994; Sano et al, 1996; Uyama et al, 1993).

Partial hepatectomy is thought to increase levels of circulating hepatotrophic substances through an unknown mechanism. This stimulates hepatocyte proliferation and signals the liver to regenerate. The standard model of partial hepatectomy in rodent models is the seventy percent partial hepatectomy described by Higgins and Anderson (Higgins and Anderson, 1931). The left and median lobes are divided by ligature or cautery and excised, leaving the right lobe. Seventy percent partial hepatectomy at the time of implantation of the hepatocyte/polymer constructs has resulted in a one to twofold increase in hepatocyte cell area by morphometric analysis in a rat model (Stein et al, 1990).

Replacement models

Efficacy of the hepatocyte-PGA constructs were analyzed in the Gunn rat and Dalmation dog models. As previously described, the Gunn rats have a deficiency in conjugation of bilirubin, resulting in elevation of unconjugated bilirubin in the serum, with an absence of conjugated bilirubin. Dalmation dogs have a deficiency in the uptake of uric acid in the liver which results in elevated serum and urine levels of uric acid (Giesecke et al, 1984). In both animal models, the implantation of normal hepatocytes on PGA sheets in the mesentery of portacaval shunted animals caused partial correction of the enzyme deficiencies for up to six weeks (Takeda et al, 1994; Uyama et al, 1993). Histology of the constructs showed viable hepatocytes around the polymer remnants surrounded by fibrovascular tissue.

SUCCESSFUL TISSUE ENGINEERING OF LIVER: SEED OR SOIL?

Despite the encouraging results of the initial hepatocyte/polymer transplantation studies and the replacement model studies, it is clear that in order to successfully tissue engineer liver tissue, further improvements in our system must be made. The focus of recent work on tissue engineering of the liver in our laboratory has been on improving the "soil" for the hepatocytes; that is, improving the environment into which the hepatocytes are implanted in order to increase survival and proliferation. Attempts to improve the soil have included optimization of the site of implantation, cotransplantion of hepatocytes with other cells, addition of sustained-released hepatotrophic growth factors, and increasing circulating hepatotrophic factors in the host animal by means of such operative manipulations as partial hepatectomy or portacaval shunt. Additionally, with the rapid advances in polymer chemistry, polymer devices have gone from simply being delivery vehicles to guiding hepatocyte attachment, growth, differentiation, and organization.

We have also attempted to improve tissue engineering of the liver by using alternative cell sources to normal adult hepatocytes. These efforts to improve the "seed" or the cell have included transplantation of fetal hepatocytes, transplantation of liver epithelial stem cells, and transplantation of genetically altered hepatocytes. In the next section we will discuss current strategies in our laboratory for improving tissue engineering of the liver.

Soil

Hepatocytes have remarkable replicative capabilities *in vivo*. The ancient Greeks knew of the ability of the liver to regenerate as illustrated by the myth of the titan Prometheus. As punishment for giving fire to man, the Greek gods bound Prometheus to a rock, and each day eagles would eat part of his liver. Every day his liver would grow back fully, only to have it eaten again by the eagles.

From the classic experiments performed in the 1930s by Higgins and Anderson, the liver was shown to be able to fully regenerate after repeated seventy percent partial hepatectomies (Higgins and Anderson, 1931). In experiments by Rhim et al on mice transgenically altered to have a functional liver deficiency, single adult hepatocytes when transplanted orthotopically have been shown to undergo at least twelve divisions (Rhim et al, 1994). Thus, when properly stimulated, the normal adult hepatocyte has the capability to undergo enough divisions to form greater than 2,000 cells. However, from the *in vitro* hepatocyte culture experience, it is clear that hepatocytes are poorly sustained in culture for prolonged periods and will undergo only minimal proliferation. Additionally, hepatocytes transplanted heterotopically without stimulation will have poor survival and proliferation. Thus, in order to take advantage of the replicative advantage of the transplanted hepatocyte, we have investigated various factors that are known to be hepatotrophic either *in vitro* or *in vivo* and applied them to improving tissue engineering of the liver.

Advances in polymers

In order to improve hepatocyte attachment, growth, and function, currently available polymers are being modified, and newer polymers are being designed. A theoretically optimal polymer would be biodegradable over a period of time that would allow the transplanted cells to synthesize their own extracellular matrix. It would have, when implanted *in vivo,* a limited inflammatory response thereby decreasing fibrosis and damage to trans-planted cells. Additionally, it would be geometrically arranged so that it would cue tissue or organ specific microscopic and macroscopic structures. In the case of the liver, it would be important to have cords of hepatocytes lined by sinusoidal endothelium with bile ductules draining the cords of hepatocytes. The polymer would be optimized for attachment of hepatocytes and would signal hepatocyte growth and differentiated function. Addition-ally, transplanted hepatocytes should have close proximity to blood supply either by rapid angiogenesis into the polymer or by having a vascular con-duit built directly into the polymer.

With rapid advances in applications of polymer chemistry to tissue engi-neering, some of these characteristics of the ideal polymer for tissue engi-neering of the liver may soon be realized. Hrkach et al (1995) have been able to attach incorporate lysine units into a backbone of PLA. Incorporation of lysine into the polymer would allow sites for insertion of functional mole-cules. Using this technique, it would theoretically be possible to attach extracellular matrix molecules such as fibronectin, laminin, collagen types I, III, and IV or growth factors such as hepatocyte growth factor (HGF) or epidermal growth factor (EGF) to optimize cell attachment, growth, and function. Lopina et al have been able to modify polyethylene oxide (PEO) radiation cross-linked hydrogels with carbohydrate moieties (Lopina et al, 1996). Hepatocyte attachment was shown to be increased when the PEO hydrogel was covalently modified with galactose. These polymers were able to support survival and differentiated function of cultured hepatocytes up to six days *in vitro.*

Mooney et al. demonstrated improved engraftment of hepatocytes using EGF incorporated into copolymer microspheres of poly-glycolic acid and poly-lactic acid (PGLA) of approximately 20 μm in diameter (Mooney et al, 1996). EGF is a known mitogen for hepatocytes *in vitro* and increases liver regeneration after partial hepatectomy *in vivo.* The microspheres were shown to degrade over the course of one month in an aqueous environment *in vitro.* As they degraded, EGF was released. The released EGF was shown to be biologically active in hepatocyte proliferation assays. The EGF/PGLA microspheres were seeded with hepatocytes onto PLA discs which were implanted onto the mesentery of porta-caval shunted rats. Control speci-mens consisted of PLA discs seeded with hepatocytes and blank PGLA microspheres. The specimens were left for four weeks in the animal and then analyzed for engraftment of hepatocytes. Specimens containing the EGF/PGLA microspheres were shown to have a twofold increase in hepa-tocyte engraftment as compared with controls. Thus, sustained-released lo-calized delivery of growth factors was shown to have a beneficial effect in heterotopic hepatocyte engraftment.

A novel method of constructing polymers with complex three-dimen-sional shapes has been developed by Cima et al PGLA in liquid state is expelled onto a powder base from a modified ink-jet printer which is computer driven to create two-dimensional polymer patterns (Griffith-Cima, 1995). The polymer solidifies upon contact with the powder base. After one layer of polymer is finished imprinting upon the powder, subsequent layers of powder and polymer are stacked on top. The powder is then removed creating empty spaces within the polymer. Using this technology, highly complex and intricate three-dimensional shapes have been created. Early work with these polymers has focused on creating cylindrical branching channels within the polymers in the attempt to recreate a macrovascular and microvascular bed. The creation of a polymer that could be directly connected to a vascular inflow and outflow would obviate the problem of diffusion of nutrients and wastes across the entire thickness of the polymer. This would allow implantation of metabolically active cells such as hepatocytes in larger numbers on thicker polymers. Also, as the technology is further refined, additional channels could be created for seeding of biliary epithelial cells in attempts to make bile ducts.

Cotransplantation studies

Cotransplantation with other cells may improve engraftment of hepatocytes in several ways. First, they may secrete hepatotrophic growth factors. Second, they may synthesize and secrete extracellular matrix molecules that increase hepatocyte survival or proliferation. Additionally, they may provide a directly beneficial cell-cell interaction to the hepatocyte.

Nonparenchymal cells

Many cells are known to synthesize substances that are hepatotrophic. Within the liver, hepatocytes (parenchymal cells) constitute about 30–50% of all cells. The remaining cells (nonparenchymal cells) include biliary epithelial cells, Ito cells, sinusoidal endothelial cells, Kuppfer cells, and epithelial cells. Of these, Ito cells, sinusoidal endothelial cells, and perhaps Kuppfer cells are thought to produce HGF. Additionally, these cells may provide structural cues or supply other factors necessary for growth or maintenance of hepatocytes, and they may be important not only in liver homeostasis but also in regeneration. For example, the sinusoidal endothelium is a highly specialized microvascular network for the liver. Unlike most other capillary beds, it expresses no von Willebrand factor, has large fenestrations, and has no basement membrane (Martinez-Hernandez and Amenda, 1995). These unique characteristics of the sinusoidal endothelium not only allow for rapid exchange of macromolecules between hepatocytes and the blood but probably contribute to maintenance of healthy hepatocytes. It has been shown that in certain diseased states of the liver, the sinusoidal endothelium becomes more "capillarized"; that is, it expresses von Willebrand factor, loses fenestrations, and deposits basement membrane (Lee et al, 1996).

The nonparenchymal cell fraction is easily isolated after a standard two-step collagenase perfusion of the liver that can also be used to isolate hepatocytes. Cotransplantation of hepatocytes with nonparenchymal cells has been shown to significantly increase hepatocyte engraftment over transplantation of hepatocytes alone (Sano et al, in 1996). This may be due to hepatotrophic growth factors released by the nonparenchymal cells. Also, the nonparenchymal cells may provide a beneficial cell-cell interaction with the hepatocytes. Finally, the addition of the nonparenchymal cells may allow ultimately for recreation of biliary ductules or sinusoidal structures.

Islets of Langerhans

Extra-hepatic cells also produce factors that stimulate hepatocyte growth. Cells in the islets of Langerhans within the pancreas produce many substances that are hepatotrophic including insulin, glucagon, and EGF. These substances are secreted into the splenic vein which drains the blood from the pancreas and carries it to the portal vein which supplies two-thirds of the blood supply to the liver. The importance of the portal blood flow for hepatic maintenance has been demonstrated in numerous studies. In experiments performed by Starzl, portal flow was shown to be more important in maintaining hepatic mass than hepatic artery blood flow (Starzl et al, 1973.

Cotransplantation of islets of Langerhans with hepatocytes on PLA discs has been shown to increase engraftment of hepatocytes when transplanted into syngenic rats as compared to transplantation of hepatocytes alone (Kaufmann et al, 1994). Large clusters of hepatocytes were centered around islets suggesting some hepatotrophic paracrine effect of the islets. Immunohistochemical staining of the specimens showed insulin, a hepatotrophic protein, localized to the islets of Langherhans.

Prevascularization

There is evidence to suggest increasing the blood supply would increase engraftment of hepatocytes within the polymers upon implantation *in vivo.* First, the liver is a very metabolically active organ, requiring an extensive blood supply. Additionally, histologic evaluation of hepatocytes seeded onto polymer and implanted *in vivo* show increased survival of cells on the periphery of the constructs, usually centered around blood vessels.

Stein et al showed that prevascularization of the polymers prior to hepatocyte implantation improved engraftment of hepatocytes (Stein et al, 1990). PVA sponges were prevascularized by implantation into the mesentery of a rat. Hepatocytes were subsequently injected into the prevascularized polymers. The prevascularized constructs had improved survival compared to controls. Takeda et al reported that eight days of prevascularization was the optimal length for prevascularization in a swine model (Takeda et al, 1995). Future studies in this area will focus on increasing vascularization by adding angiogenic factors to the polymers or by cotransplantation with endothelial cells.

Seed

The major difficulty with tissue engineering of liver tissue is the poor survival and proliferation of adult hepatocytes in *in vitro* culture conditions and when heterotopically transplanted. Thus, not only is *in vitro* amplification of cell number not possible, there is limited cell survival and proliferation *in vivo* after transplantation. Because of this, we have explored alternative cell types that may be able to perform all or some of the functions as adult hepatocytes.

Fetal hepatocytes

Fetal hepatocytes are known to have a greater potential to proliferate *in vitro* and *in vivo*. Cusick et al compared transplantation of fetal and adult hepatocytes on PLA discs in portacaval shunted rats. Fetal hepatocytes were isolated from livers of animals of between 15 and 17 days of gestation. Constructs containing either fetal or adult hepatocytes were implanted into adult syngenic rats. Constructs containing fetal hepatocytes showed a four to elevenfold increase in hepatocyte cell area compared to those containing adult hepatocytes by morphometric analysis (Cusick et al, 1995).

Liver stem cells

Liver epithelial stem cells, also referred to as oval cells, have been thought by some to be pluripotent liver cells, able to differentiate into either biliary epithelial cells or hepatocytes. Oval cells are easily propagable in culture and thus would be a potential source of hepatocytes for tissue engineering of liver. Perhaps the best-characterized oval cell line is the WB-344 cell line isolated by Ming-Sound et al. The WB-344 cell line has been demonstrated to have some phenotypic characteristics of biliary epithelial cells and some phenotypic characteristics of hepatocytes (Ming-Sound et al, 1984). Our laboratory is currently investigating oval cell implantation on polymer.

Genetically altered hepatocytes

Another method to improve tissue engineering of the liver would be genetically altering normal adult hepatocytes. Shiota et al have described creation of a transgenic strain of mice with increased expression of HGF in hepatocytes via an albumin promoter (Shiota et al, 1994). Our laboratory is currently investigating the effect of transplantation of HGF-transgenic hepatocytes on PLA polymers as compared with wild-type hepatocytes. Additionally, in order to determine the effect of the recipient environment, polymer/hepatocyte constructs implanted into HGF-transgenic mice will be compared to constructs implanted into wild-type mice. Preliminary results show an increased engraftment of HGF-transgenic hepatocytes compared to wild-type hepatocytes.

CONCLUSION

End-stage liver disease is a morbid and often lethal disease. Current treatment modalities are limited. The only accepted definitive treatment is liver transplantation which has a high associated morbidity. Additionally, the number of donor livers is extremely limited, and the shortage of donor organs is ever worsening. Alternative treatments could save many lives and prevent many deleterious side-effects in patients who at present receive only supportive therapy while awaiting a donor organ. One of the promising approaches to hepatic replacement is tissue engineering of the liver. The advances in clinical medicine, molecular and cellular biology, and polymer chemistry present the tissue engineer with powerful tools for advances in constructing a tissue engineered liver.

Acknowledgments The authors would like to thank Jane Landis for her usual expertise in preparation of the manuscripts, and Robert A. Cusick, Joerg-Matthias Pollok, Hirofumi Utsunomiya, and Kerilyn Nobuhara for their critical review.

References

Aiken J, Cima L, Schloo B, Mooney D, Johnson L, Langer R, Vacanti JP (1990): Studies in rat liver perfusion for optimum harvest of hepatocytes. *J Pediatr Surg* 25:140–145

American Liver Foundation (1988): *Vital Statistics of the United States,* Vol 2, Part A. In: Langer R, Vacanti J (1994): Tissue engineering. *Science* 260:920–926

Broelsch CE, Burdelski M, Rogiers X, Gundlach M, Knoefel WT, Langwieler T, Fischer L, Latta A, Hellwege H, Schulte FJ, Schmiegel W, Sterneck M, Greten H,

Kuechler T, Krupski G, Loeliger C, Kuehnl P, Pothmann W, Schultz-Am-Esch J (1994): Living donor for liver transplantation. *Hepatology* 20:49S–55S

Bucher NLR (1991): Liver regeneration: An overview. *J Gastroent Hepatol* 6:615–624

Cusick RA, Sano K, Lee H, Pollok JM, Mooney D, Langer R, Vacanti JP (1995): Heterotopic fetal rat hepatocyte transplantation on biodegradable polymers. *Surgical Forum* 46:658–661

Demetriou AA, Whiting JF, Feldman D, Levenson SM, Chowdury NR, Moscioni AD, Kram M, Chowdury JR (1986): Replacement of liver function in rats by transplantation of microcarrier-attached hepatocytes. *Science* 233:1190–1192

Emond JC, Whitington PF, Thistlethwaite JR, Cherqui D, Alonso EA, Woodle IS, Vogelbach P, Busse-Henry SM, Zuker AR, Broelsch CE (1990): Transplantation of two patients with one liver. *Ann Surg* 212(1):14–22

Gerlach JC, Schnoy N, Encke J, Smith MD, Muller C, Neuhaus P (1995): Improved hepatocyte in vitro maintenance in a culture model with woven multicompartment capillary systems: Electron microscopy studies. *Hepatology* 22:546–552

Giesecke DD, Tiemeyer W (1984): Defect of uric acid in dalmation dog liver. *Experientia* 40:1415–1416

Griffith-Cima LN (1995): Personal communications

Grossman M, Roper SE, Kozarsky K, Stein EA, Engelhardt JF, Muller D, Luppien PJ, Wilson JM (1994): Successful ex vivo gene therapy directed to liver in a patient with familial hypercholesterolemia. *Nature Genetics* 6:3355–3341

Higgins GM, Anderson RM (1931): Experimental pathology of the liver. I. Restoration of the liver of the white rat following partial surgical removal. *Arch Pathol* 12:186–202

Hrkach JS, Ou J, Lotan N, Langer R (1995): Synthesis of poly (L-lactic acid-co-lysine) graft copolymers. *Macromolecules* 28:4736–4739

Johnson LB, Aiken J, Mooney D, Schloo BL, Griffith-Cima L, Langer R, Vacanti JP (1994): The mesentery as a laminated vascular bed for hepatocyte transplantation. *Cell Transplant* 3(4): 273–281

Kaufmann PM, Sano K, Uyama S, Schloo B, Vacanti JP (1994): Heterotopic hepatocyte transplantation using three-dimensional polymers: Evaluation of the stimulatory effects by portacaval shunt or islet cell cotransplantation. *Transplant Proc* 26:3343–3345

Langer R, Vacanti J (1993): Tissue engineering. *Science* 260:920–926

Lee H, Cusick RA, Khachadourian H, Vacanti JP, Perez-Atayde A (1996): Pathologic alterations in the hepatic sinusoids may contribute to the progression of cirrhosis in EHBA. Presented at the American Pediatric Surgical Association. San Diego, CA: May 19–22

Lopina ST, Wu G, Merrill EW, Griffith-Cima L (1996): Hepatocyte culture on carbohydrate-modified star polyethylene oxide hydrogels. *Biomaterials:* in press

Martinez-Hernandez A, Amenta PS. (1995): The extracellular matrix in hepatic regeneration. *FASEB J* 9:1401–1410

Matas AJ, Sutherland DER, Steffes MW, Mauer SM, Lowe A, Simmons RL, Najarian JS (1976): Hepatocellular transplantation for metabolic deficiencies: Decrease of plasma billirubin in Gunn rats. *Science* 192:892–894

Ming-Sound T, Smith JD, Nelson KG, Grisham JW (1984): A diploid epithelial cell line from normal adult rat liver with phenotypic properties of oval cells. *Exp Cell Res* 154:38–52

Mooney DJ, Cima L, Langer R, Johnson L, Hansen LK, Ingber DE, Vacanti JP (1992): Principles of tissue engineering and reconstruction using polymer cell constructs. *Mat Res Soc Symp Proc* 252:345–352

Mooney DJ, Kaufmann PM, Sano K, McNamara KM, Vacanti JP, Langer R (1994): Transplantation of hepatocytes using porous, biodegradable sponges. *Transplant Proc* 26:3425–3426

Mooney DJ, Kaufmann PM, Sano K, Schwendeman SP, McNamara K, Schloo B, Vacanti JP, Langer R (1996): Localized delivery of EGF improves the survival of transplanted hepatocytes. *Biotech Bioeng* 50:422–429

Onoder K, Ebata H, Sawa M, Katoh K, Kasai S, Mito M, Nozawa M (1992): Comparative effects of hepatocellular transplantation in the spleen portal vein, or peritoneal cavity in congenitally ascorbic acid biosynthetic enzyme-deficient rats. *Transplant Proc* 24(6):3006–3008

Ponder KP, Gupta S, Leland F, Darlington G, Finegold M, DeMayo J, Ledley FD, Chowdury JR, Woo SL (1991): Mouse hepatocytes migrate to liver parenchyma and function indefintely after intrasplenic transplantation. *Proc Natl Acad Sci USA* 88(4):1217–1221

Rhim JA, Sandgren EP, Degen JL, Palmiter RD, Brinster RL (1994): Replacement of diseased mouse liver by hepatic cell transplantation. *Science* 263:1149–1152

Rozga J, Podesta L, Lepage E, Morsiani E, Moscioni AD, Hoffmann A, Sher L, Villamil F, Woolf G, McGrath M, Kong L, Rosen H, Lanman T, Vierling J, Makowka L, Demetriou AA (1994): A bioartificial liver to treat severe acute liver failure. *Ann Surg* 219(5):538–546

Sano K, Cusick RA, Lee H, Pollok JM, Kaufmann PM, Uyama S, Mooney D, Langer R, Vacanti JP (1996): Regenerative signals for heterotopic hepatocyte transplantation. *Transplant Proc:* in press

Seglen PO (1976): Preparation of isolated rat liver cells. *Meth Cell Biol* 13:29–83

Shiota G, Wang TC, Nakamura T, Schmidt EV (1994): Hepatocyte growth factor in transgenic mice: Effects on hepatocyte growth, liver regeneration and gene expression. *Hepatology* 19:962–972

Starzl TE, Francavilla A, Halgrimson CG, Francavilla FR, Porter KA, Brown TH, Putnam CW (1973): The origin, hormonal nature, and action of hepatotrophic substances in portal venous blood. *Surg Gynecol Obstet* 137:179–198

Starzl TE, Fung J, Tzakis A, Todo S, Demetris AJ, Marino IR, Doyle H, Zeevi A, Warty V, Michaels M, Kusne S, Rudent WA, Trucco M (1993): Baboon-to-human liver transplantation. *Lancet* 341:65–71

Stein JE, Asonuma K, Gilbert JC, Schloo B, Ingber D, Vacanti JP (1990): Hepatocyte transplantation using prevascularized porous polymers and hepatotrophic stimulation. Presented at American Pediatric Surgical Association meeting. Lake Buena Vista, FL: May 15–18

Sussman NL, Chong MG, Koussayer T, He D, Shang TA, Whisennand HH, Kelly JH (1992): Reversal of fulminant hepatic failure using an extracorporeal liver assist device. *Hepatology* 16:60–65

Takeda T, Kim TH, Lee SK, Davis J, Vacanti JP (1994): Hepatocyte transplantation in biodegradable polymer scaffolds using the dalmation dog model of hyperuricosuria. Presented at XVth Congress of the Transplantation Society. Koyoto, Japan: Aug 28–Sep 2

Takeda T, Murphy S, Uyama S, Organ GM, Schloo BL, Vacanti JP (1995): Hepatocyte transplantation in swine using prevascularized polyvinyl alcohol sponges. *Tissue Eng* 1(3) 253–259

US Scientific Registry for Organ Transplantation and the Organ Procurement and Transplantation Network (1990): *Annual Report.* In: Langer R, Vacanti J (1994): Tissue engineering. *Science* 260:920–926

Uyama S, Kaufmann PM, Takeda T, Vacanti JP (1993): Delivery of whole liver-equivalent hepatocyte mass using polymer devices and hepatotrophic stimulation. *Transplantation* 55:932–935

Vacanti JP, Morse MA, Saltzman M, Domb AJ, Perez-Atayde A, Langer R (1988): Selective cell transplantation using bioabsorbable artificial polymers as matrices. *J Pediatr Surg* 23:3–9

Yarmush ML, Toner M, Dunn JCY, Rotem A, Hubel A, Tompkins RG (1992): Hepatic tissue engineering: Development of critical technologies. *Ann NY Acad Sci* 665:238–252

INDEX